MATERIAL IDENTIFICATION USING
MIXED NUMERICAL EXPERIMENTAL METHODS

Material Identification Using Mixed Numerical Experimental Methods

Proceedings of the EUROMECH Colloquium held in Kerkrade, The Netherlands, 7-9 April 1997

Edited by

H. Sol
Faculty of Applied Sciences,
Free University of Brussels,
Belgium

and

C.W.J. Oomens
Mechanical Engineering Department,
Eindhoven University of Technology,
The Netherlands

KLUWER ACADEMIC PUBLISHERS
DORDRECHT / BOSTON / LONDON

A C.I.P. Catalogue record for this book is available from the Library of Congress.

ISBN-13:978-94-010-7158-1 e-ISBN-13:978-94-009-1471-1
DOI:10.1007/978-94-009-1471-1

Published by Kluwer Academic Publishers,
P.O. Box 17, 3300 AA Dordrecht, The Netherlands.

Sold and distributed in the U.S.A. and Canada
by Kluwer Academic Publishers,
101 Philip Drive, Norwell, MA 02061, U.S.A.

In all other countries, sold and distributed
by Kluwer Academic Publishers,
P.O. Box 322, 3300 AH Dordrecht, The Netherlands.

EUROMECH 357 has been sponsored by the Technology Foundation (STW)

Printed on acid-free paper

PREFACE

Euromech 357 took place in the nice authentic monastery 'Rolduc' in Kerkrade, Holland. The objective was to bring together researchers to present their latest advancements in the relatively new domain of Material properties identification by Mixed Numerical Experimental Methods (MMM).
MMM are a modern and increasingly powerful way to determine the values of unknown parameters in a numerical model by observations made on real physical test structures. Starting from the measurement of output values (like displacements, stresses, velocities, vibrations,..) of the real physical test structure, MMM try to update parameters in the numerical model in such a way that the computed observations match the experimental observations. It was shown clearly at the colloquium that the combined use of numerical analysing tools and sophisticated measurement techniques has created an extra degree of freedom for the design of experiments and has led to new approaches for material characterisation. The colloquium was attended by 39 participants coming from 12 different European countries.
We had 23 oral presentations on the different topics of the above mentioned sessions. Each presentation was followed by an intensive discussion. Due to the informal atmosphere and the limited number of participants, the discussions were very lively and fruitful. The opportunity to continue to discuss common problems after dinner in a reserved place was also offered to the Euromech participants. This opportunity and the fact that the monastery was isolated (nobody could escape!) created an excellent platform for discussions and personal contacts.
The second day, a poster session was organised in the evening. The goal of the posters was to serve as a starting point for discussions for those participants who could not give an oral presentation. (The number of presentations was limited in order to create much time for discussions).
As a result of the final discussion the last day, a permanent research group MMM was created for which 16 participants offered active co-operation. The main goals of the Research group MMM are to offer a permanent discussion platform, to write lecture notes on some important common topics and to organise workshops (possibly an other Euromech colloquium) in the future.
The common important scientific conclusion of the colloquium was that the necessary demand for success of MMM is a very accurate numerical model of the experimental set-up and a sufficiently accurate constitutive model to describe the material behaviour. Moreover, the experimental observations have to contain sufficient information to be enable to estimate the parameters. The question is which criterion can be used for this decision and how observability can be improved. Other topics were the weighting of different types and numbers of observations and the way to perform a sensitivity analysis of the experimental set-up. The research group MMM will work out an in depth analysis of these statements.

Hugo Sol, Cees Oomens

CONTENTS

SCIENTIFIC ADVISORY COMMITTEE

P. Hamelin
Université Claude Bernard, France

M.E.H. Hendriks
T.N.O.Bouw, Delft, The Netherlands

W. Heylen
Katholieke Universiteit Leuven, Belgium

J.D. Janssen
Eindhoven University of Technology, The Netherlands

H.E.H. Meijer
Eindhoven University of Technology, The Netherlands

P. Pedersen
Dept. Solid Mechanics, DTU, Denemarken

W.P. de Wilde
Free University of Brussels, Belgium

ORGANIZING COMMITTEE

H.Sol
Free University of Brussels, Belgium

C.W.J.Oomens
Eindhoven University of Technology, The Netherlands

M. H.W.van Boxtel-Rieken
Eindhoven University of Technology, The Netherlands

IDENTIFICATION OF THE COMPLEX MODULI OF THIN FIBRE REINFORCED POLYMER PLATES USING MEASURED MODAL PARAMETERS

J. DE VISSCHER, H. SOL, W.P. DE WILDE
Vrije Universiteit Brussel, Dienst Analyse van structuren
Pleinlaan 2, 1050 Brussel
J. VANTOMME
Koninklijke Militaire School, Departement Bouwkunde
Renaissancelaan 30, 1000 Brussel

Abstract. Complex moduli are used to characterize the dynamic material behaviour of fibre reinforced polymers (FRP). This paper presents an efficient method to identify the in-plane complex moduli of thin orthotropic FRP plates, based on a mixed numerical experimental approach. The required experimental data are the modal parameters of a set of specific eigenmodes of a plate specimen. In combination with a numerical model for the calculation of the modal parameters of the specimen, all the independent complex moduli of the material are derived.

1. Introduction

The challenge of designing with fibre reinforced polymers (FRP) is that the material properties can be "tailored" to meet the specific requirements of the given application. For this purpose, composite laminate analysis programs are available to choose the design parameters in a very early phase. Yet, the need for experimental material identification techniques to verify the properties of the final laminate in a reliable, fast and preferably non-destructive way will always exist: laminate analysis programs are based on many theoretical assumptions, often unreliable input data and they can not account for all production variables that influence the resulting material properties.

For FRP, the number of material parameters required to describe the mechanical behaviour is relatively large, due to the anisotropy (fibre reinforcement) and the viscoelastic nature of the materials. The importance of this viscoelasticity should not be ignored, since FRP are most frequently used in dynamic applications where requirements such as low weight, high stiffness and high material damping prevail. Hence, traditional experimental methods, such as static (and dynamic) tension, bending and torsion tests on beam-shaped specimens, are too cumbersome and time-consuming. Moreover, there are some drawbacks to the use of beam specimens, specifically for testing FRP:

- due to the inevitable heterogeneity of FRP, results may vary from beam to beam specimen, even when they are cut from the same plate;
- the smaller the width of the specimens, the more the results are susceptible to edge defects;

H. Sol and C.W.J. Oomens (eds.), Material Identification Using Mixed Numerical Experimental Methods, 1-9.

- the stress state in beam specimens is uniaxial, so that coupling moduli can not be measured (unless biaxial strain-gauges are used, which is once again expensive and time-consuming).

In view of all these problems, the idea came up at the Free University of Brussels to develop a material identification technique for FRP, which is based on *dynamic quantities*, measured on *plate specimens*.

Sol [1] worked out the idea by developing a method for the identification of the elastic moduli of orthotropic thin FRP plates. He proposed to measure a set of eigenfrequencies: eigenfrequencies are higly sensitive to the material stiffness, while they can be measured quickly, accurately and with simple equipment. He developed a numerical model to establish the relation between the desired quantities (elastic moduli) and the measured quantities (resonant frequencies). This mixed numerical experimental technique was named the *Resonalyser*.

Later on, research concentrated on the identification of FRP material damping, which is mainly a consequence of the viscoelastic nature of the phases and the fibre/matrix interaction. Therefore, the modal damping ratios associated to the set of resonant frequencies are also measured and the same numerical model is used to derive the material damping properties [2]. The Resonalyser is thus capable of identifying all the relevant material properties for describing the dynamic behaviour (e.g., to provide input parameters for Finite Element calculations).

2. Theoretical concepts

2.1. COMPLEX MODULI

Linear viscoelasticity is generally accepted as a good theory to describe the behaviour of FRP for small deformations. The Boltzmann superposition integral thus allows to express the stress as a function of the strain history [Christensen]:

$$\sigma_i(t) = \int_{-\infty}^{t} C_{ij}(t-\tau)\frac{\partial \varepsilon_j}{\partial \tau}d\tau \tag{1}$$

where Cij(t) is the tensor of relaxation functions.
Expressing the strain as a superposition of harmonic strain components:

$$\varepsilon_j(t) = \varepsilon_{jo}e^{i\omega t} \tag{2}$$

and introducing the variable η=t-τ, yields the stress-strain relation for steady state vibrations:

$$\begin{aligned}\sigma_i(t) &= \omega\,[\int_0^\infty C_{ij}(\eta)\,\sin\omega\eta\,d\eta + i\int_0^\infty C_{ij}(\eta)\,\cos\omega\eta\,d\eta\,]\,\varepsilon_{jo}\,e^{i\omega t}\\ &= [C_{ij}'(\omega) + i\,C_{ij}''(\omega)]\,\varepsilon_{jo}\,e^{i\omega t}\end{aligned} \tag{3}$$

or:

$$\sigma_i^* = C_{ij}^*(\omega)\varepsilon_j^* \tag{4}$$

where $C_{ij}^*(\omega) = i\omega\, F(C_{ij}(t))$ and $F(C_{ij}(t))$ are the Fourier Transformed relaxation functions.

Relation (4) is similar to the stress strain-relation of an elastic material, except that the components of the stiffness matrix are complex instead of real. The real parts are called the *elastic* or *storage* moduli, since they determine the potential energy stored within the deformed material. The imaginary parts are called the *loss* moduli, since they determine the amount of energy dissipated or lost under deformation. The loss moduli are thus the material properties that characterize material damping.

Relation (4) also shows that when a linear viscoelastic material is subjected to steady state vibrations, a phase shift $\delta(C_{ij})$ exists between the various stress and strain components (see figure 1).

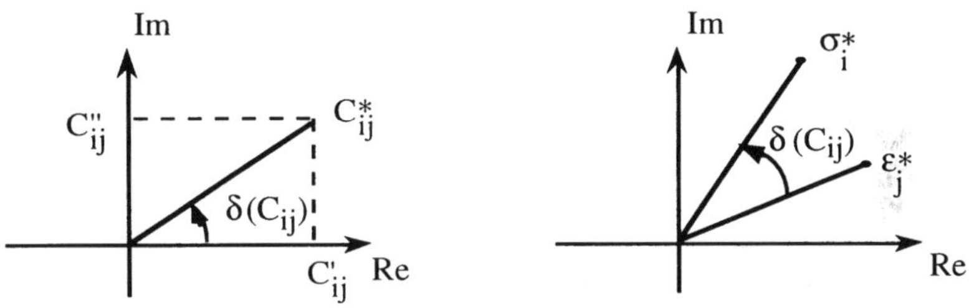

Figure 1. Graphical representation of the phase shift $\delta(C_{ij})$

For thin orthotropic FRP plates, the complex stiffness matrix can be written in terms of 4 independent complex engineering constants: the 2 Young's moduli E_1^* and E_2^*, the shear modulus G_{12}^* and Poisson's ratio ν_{12}^*:

$$\begin{Bmatrix} \varepsilon_1^* \\ \varepsilon_2^* \\ \gamma_6^* \end{Bmatrix} = \begin{bmatrix} 1/E_1^* & -\nu_{21}^*/E_2^* & 0 \\ -\nu_{12}^*/E_1^* & 1/E_2^* & 0 \\ 0 & 0 & 1/G_{12}^* \end{bmatrix} \begin{Bmatrix} \sigma_1^* \\ \sigma_2^* \\ \sigma_6^* \end{Bmatrix} \tag{5}$$

2.2. IDENTIFICATION OF THE ELASTIC MODULI

The numerical model developed for calculating the free vibrations of thin anisotropic plates is based on the Ritz method: the exact displacement field is approximated by a linear combination of a finite number of selected functions. Sol [1] used Lagrange polynomials, thus obtaining convergence of the numerical solution with only 49 degrees of freedom. The eigenfrequencies and eigenmodes of a thin anisotropic plate can thus be calculated numerically by solving the eigenvalue problem:

$$([K'] - \omega^2\, [M])\, \{\Phi\} = \{0\} \tag{6}$$

where the stiffness matrix [K'] is assembled using the elastic moduli. Comparing the numerically calculated eigenfrequencies to the associated set of measured eigenfrequencies allows to update the values of the elastic moduli used to assemble [K']. This is repeated until measured and calculated eigenfrequencies are sufficiently close. A

Bayesian parameter estimation technique is applied to calculate the updated elastic moduli in each iteration cycle. Figure 2 shows the general scheme of the procedure.

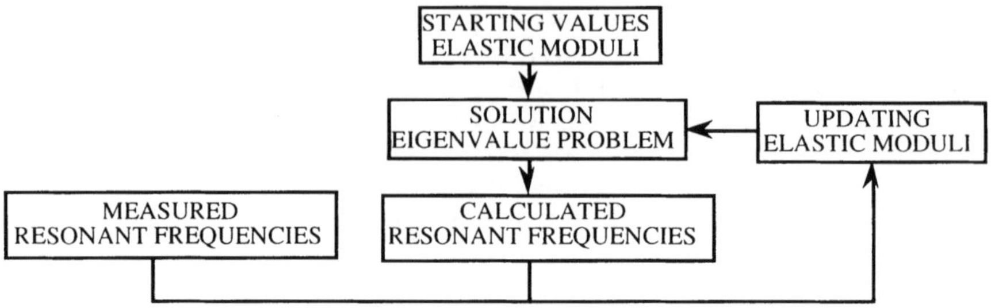

Figure 2. Procedure for identifying the elastic moduli

Remark: For a viscoelastic material, the complex moduli $C_{ij}*$ should be used to assemble the stiffness matrix. This would lead to a complex eigenvalue problem, yielding complex eigenfrequencies of which the real parts are the resonant frequencies and the ratios of the imaginary parts to the real parts are the modal damping ratios. The neglecting of the loss moduli in the procedure described above is justified by the fact that for FRP, the loss moduli are very small compared to the elastic moduli. Typical values for the ratio Cij"/Cij' range from 10^{-2} to 10^{-4}.

2.3. IDENTIFICATION OF THE LOSS MODULI

As mentioned in section 2.2, modal damping ratios can be calculated numerically by solving the complex eigenvalue problem. Similar to the mixed numerical experimental procedure for identifying the elastic moduli, one could thus iteratively update the loss moduli in the numerical model until the calculated and the measured damping ratios are sufficiently close. Based on the fact that the loss moduli are small, another approach is followed which is computationally far more simple and fast: the energy approach [4]. When modal damping is small, the modal damping ratio of an eigenmode of a structure can be written in terms of 2 energy quantities: the modal strain energy P and the modal damping energy D:

$$\xi = \frac{D}{4\pi P} \tag{7}$$

These energies can be expressed in terms of the complex moduli (automatic summation over the repeated indices):
Modal strain energy:

$$P = \frac{1}{2}\int_V C_{ij}'\varepsilon_{jo}\varepsilon_{io}dv \tag{8}$$

Modal damping energy:

$$D = \frac{1}{2}\int_V \pi C_{ij}''\varepsilon_{jo}\varepsilon_{io}dv \tag{9}$$

For the case of thin orthotropic plates, eq. (7) is rewritten as:

$$\xi = \frac{1}{2}[W_{11}tg\delta(C_{11}) + W_{22}tg\delta(C_{22}) + W_{12}tg\delta(C_{12}) + W_{66}tg\delta(C_{66})] \qquad (10)$$

where, for example $W_{11} = \frac{P_{11}}{P} = \frac{1}{2}\frac{\int_V C_{11}'\varepsilon_{1o}\varepsilon_{1o}dv}{P}$

W_{ij} is the ratio of the modal strain energy associated to the cycle σ_i-ε_j and the total modal strain energy. Equation (10) thus means that the contribution of a specific loss tangent to the modal damping ratio depends on the contribution of the associated strain energy to the modal strain energy.

To summarize, a linear relation is obtained between the modal damping ratio of an eigenmode and the loss tangents of the 4 complex moduli. Obviously, this approach is very convenient, since the linearity of the relation implies that the loss tangents can be obtained in only one step, whereas the identification of the elastic moduli requires several iterations. The coefficients W_{ij} in (10) are calculated using the same numerical model and the previously identified values of the elastic moduli [2].

2.4. FREQUENCY DEPENDENCE OF THE COMPLEX MODULI

The elastic moduli are practically independent of the frequency, since the polymer matrix materials of commonly used FRP are in the glassy state. This is not the case for the loss moduli, which are known to exhibit significant frequency dependence, even outside the transition regions of the polymer. When such frequency dependence is observed, the modal damping ratios need to be measured as functions of frequency. Practically, this is achieved by varying the dimensions of the test specimens and by curve-fitting the measured modal damping ratios as functions of frequency.

3. Practical procedure

3.1. SPECIMENS AND EIGENMODES

The *Resonalyser* method makes use of a set of experimental resonant frequencies and associated modal damping ratios, measured on plate specimens with free boundary conditions. The plate dimensions are arbitrary, but the length to width ratio should satisfy the relation:

$$\frac{l}{w} = \sqrt[4]{\frac{E_1'}{E_2'}} \qquad (11)$$

The first 3 eigenmodes of such a plate, named a *Poisson plate,* are shown in figure 3.

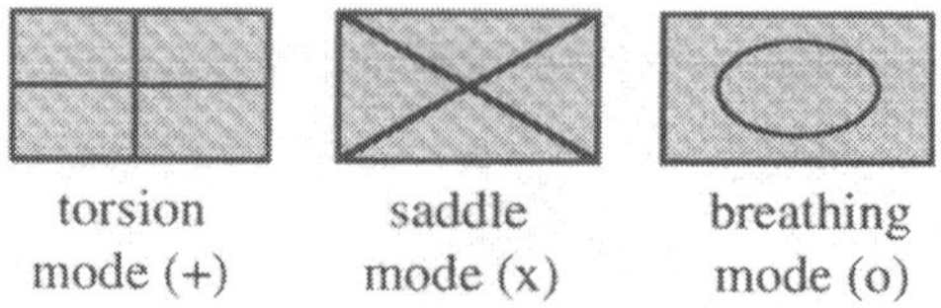

Figure 3. First 3 eigenmodes of a *Poisson plate*

The main reason for using a Poisson plate is that the (x)-mode and the (o)-mode are the modes that are the most sensitive to the coupling modulus C_{12}^* (which yields ν_{12}^*). Using arbitrary dimensions, it could be that the eigenmodes are insufficiently sensitive to Poisson's ratio, resulting in an inaccurate value. To determine the ratio in equation (11), prior knowledge of E_1' and E_2' is required. Young's moduli can be calculated analytically and with sufficient accuracy from the first resonant frequencies of beam specimens. At the same time, these values are used as starting values for the iterative procedure (figure 2). The first eigenmodes of the beam specimens, combined with the 3 first eigenmodes of the Poisson plate, are the 5 eigenmodes that are used to identify the 4 independent complex moduli.

3.2. EXPERIMENTAL SET-UP

Figure 4 shows the equipment used for the measurement of the resonant frequencies and modal damping ratios.

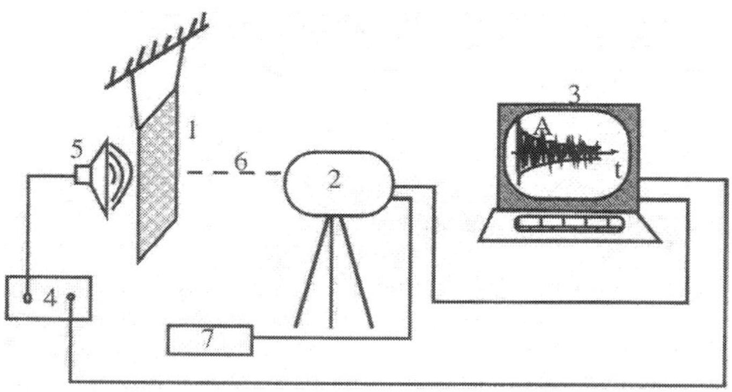

1: test plate
2: laser vibrometer
3: Personal Computer
4: Power Amplifier
5: loudspeaker (or disconnectable shaker)
6: laser beam
7: power supply laser

Figure 4. Experimental set-up

To measure the resonant frequencies, the specimen is excited by a wide spectrum signal, provided by a hammer impact or a multi-sine signal (fed through a loudspeaker or a disconnectable shaker). The specimen's response is measured by an accelerometer or a laser vibrometer and sampled by the P.C. The signal is transformed to the frequency domain, where the peaks are automatically identified as resonant frequencies.

To measure the modal damping ratios, the eigenmodes are excited one by one by a harmonic signal, either through a loudspeaker or a shaker. The harmonically decaying vibrations are measured using the laser vibrometer and sampled from the moment the excitation is stopped. Curve-fitting yields the modal damping ratio of the considered eigenmode.

4. Example

The complete identification procedure for the case of a $[0°, 90°]_{3S}$ carbon/epoxy laminated plate is shown. Carbon/epoxy composites are critical test cases because of 2 reasons. First of all, the damping of carbon/epoxy composites is extremely small: the high stiffness of the carbon fibres implies that most of the strain energy is concentrated in the fibres, which have a low damping capacity. The second difficulty is that the damping of epoxy is frequency dependent. Hence, identification as function of frequency is necessary.

Table 1 shows the dimensions, mass and resonant frequencies of the beams and the Poisson plate. Beam 1 is the beam specimen for which the fibres in the outer layers are oriented longitudinally. Since the considered modeshape is the first bending mode, beam 1 is stiffer than beam 2 ($E_1'>E_2'$). The data from table 1 allow to identify the elastic moduli, which are assumed to be frequency independent. The results are shown in table 2.

TABLE 1. Dimensions, mass and resonant frequencies of the test specimens

	length (cm)	width (cm)	thickness (mm)	mass (g)	f1 (Hz)	f2 (Hz)	f3 (Hz)
beam 1	30.03	1.607	1.77	12.69	140.0		
beam 2	28.41	1.647	1.74	12.13	123.0		
plate	29.655	26.715	1.78	210.34	38.13	139.35	145.89

TABLE 2. Elastic moduli and engineering constants

C_{11}' (N/m^2)	C_{22}' (N/m^2)	C_{12}' (N/m^2)	C_{66}' (N/m^2)	E_1' (GPa)	E_2' (GPa)	ν_{12}'	G_{12}' (GPa)
71.17 E9	45.64 E9	3.64 E9	3.66 E9	70.88	45.46	0.08	3.65

Graphs 1 and 2 show the modal damping ratios as functions of frequency. The modal damping ratio of beam 1 is smaller than that of beam 2. This is once again explained on the basis of energy considerations. The outer layers of the beams contribute most to the modal strain energy of the first bending mode. For beam 2, the fibres in the outer layers are oriented transversely on the stressed direction, so that the modal damping ratio is more determined by the epoxy matrix (wich has a higher damping than the carbon fibres). For the Poisson plate, it is observed that the modal damping ratios of the (x)- and (o)-modes are small compared to the modal damping ratio of the (+)-mode. That is because the (+)-mode is governed by shear deformation, for which the strain energy is concentrated in the matrix, while the (x)- and (o)-modes are governed by normal deformations. This is shown numerically in table 3, which shows the contribution coefficients W_{ij} for the different eigenmodes.

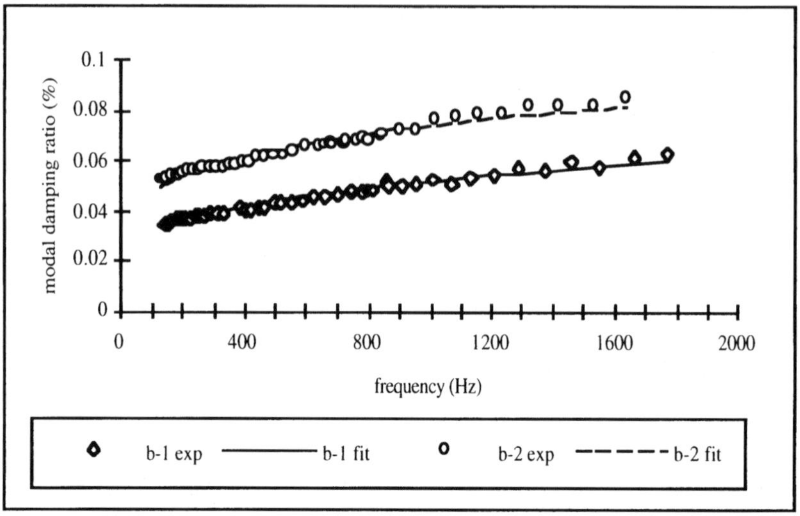

Graph 1. Modal damping ratios of the beam specimens as functions of frequency

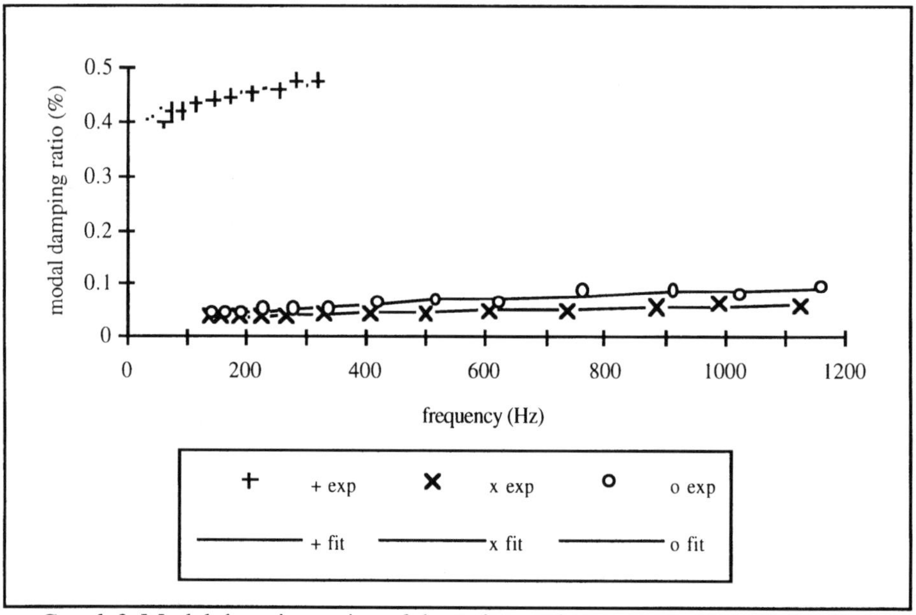

Graph 2. Modal damping ratios of the Poisson plate as functions of frequency

TABLE 3. Contribution coefficients for the 3 first eigenmodes of the Poisson plate

	fres (Hz)	W_{11} (%)	W_{22} (%)	W_{12} (%)	W_{66} (%)
(+)-mode	38.13	1.32	1.35	0.02	97.31
(x)-mode	139.35	37.54	67.15	-4.69	0
(o)-mode	145.89	63.20	33.10	3.65	0.05

The loss moduli can now be identified at discrete frequency steps (see graph 3).

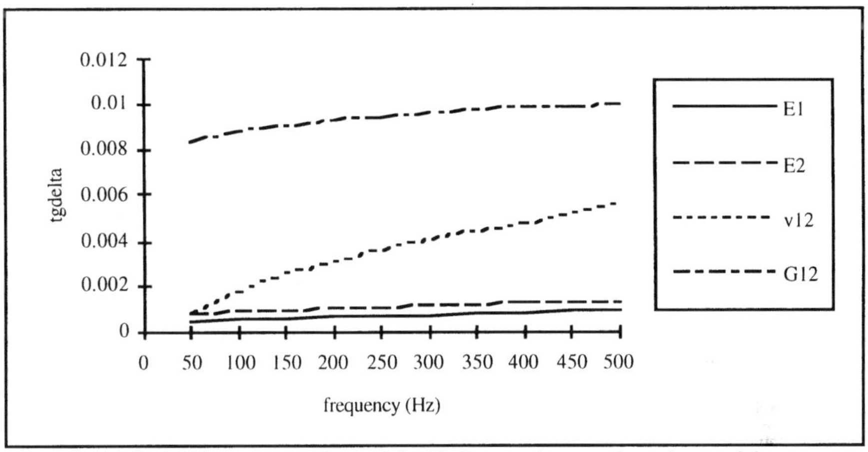

Graph 3. Modal damping ratios of the Poisson plate as functions of frequency

5. Conclusions

A method was developed for the identification of the complex moduli of FRP, where the experimental data are vibration data measured on plate specimens. Therefore, a mixed numerical experimental approach was necessary. The technique is useful for many purposes, but the most important application is the identification of material properties for the calculation of the dynamic behaviour of composite structures (e.g. using the Finite Element Method). As composite structures are generally assemblies of plates and shells, the stress states in such structures are multixial, requiring not only the knowledge of Young's moduli but also of the shear modulus and the coupling modulus (determined by ν_{12}^*). All of these data are identified simultaneously using the *Resonalyser* technique.

References

1. Sol, H.: *Identification of anisotropic plate rigidities using free vibration data,* PhD dissertation, Vrije Universiteit Brussel, 1986.
2. De Visscher, J.: *Identification of the complex stiffness matrix of orthotropic materials by a mixed numerical experimental method,* PhD dissertation, Vrije Universiteit Brussel, 1995.
3. Christensen, R.M.: *Theory of Viscoelasticity, an introduction*, Academic Press, 1971.
4. Ungar, E.E. and Kerwin, E.M.: *Loss factors of viscoelastic systems in terms of energy concepts,* J. Acoustical Society of America **34**, 7 (1962), 954-957.

CHARACTERIZATION OF INTERPHASE CONDITIONS IN COMPOSITE MATERIALS

P.F.M. MEURS, P.J.G. SCHREURS and T. PEIJS
Eindhoven University of Technology,
Centre for Polymers and Composites,
P.O. Box 513, 5600 MB Eindhoven, The Netherlands

Abstract: An experimental-numerical method is presented, which can be used for the characterization of microphenomena in composite materials. The method is based on the measurement of displacements of material points on the micron-scale, and numerical simulations of the experiments to find material constants using an iterative procedure. Displacements of a grid of markers, created with the electron beam of a Scanning Electron Microscope, were measured during in-situ experiments in the SEM. The position of the boundary markers were used to define the geometry of the model, and their displacements were used as boundary conditions. The method has been used to estimate mechanical properties of interphases and the determination of the interfacial normal strength (INS). Based on the presented results, it can be concluded that the presented method is a suitable tool for the characterization of microphenomena in composite materials.

1 Introduction

The mechanical properties of a composite material are determined by the properties of its constituents. Due to the strong inhomogeneity of these materials, i.e. the high stiffness ratio between fibres and matrix, and the inhomogeneous fibre distribution in a composite, the mechanical behaviour is strongly influenced by phenomena occurring at a scale of microns. For example, during transverse loading of unidirectional fibre-reinforced composites, local stress and strain concentrations around fibres determine the global macroscopic behaviour (1). Increasing the transverse load will lead to initiation of microcracks at the fibre/matrix interface, followed by the propagation of these cracks and macroscopic failure of the composite structure. To be able to analyze these phenomena, constitutive relations for the mechanical properties of the constituents are needed. Therefore new and better methods are needed to determine these properties. Often, such methods are based on the use of field quantities, such as displacements and strains.

In recent years, increased attention to microphenomena in composite materials resulted in an increased use of microscopic techniques in a wide range of experiments. For example, optical microscopes in combination with miniature tensile stages are used to study fibre breaking and the growth of debonding cracks along fibres in single and multiple fibre fragmentation tests (2; 3). Daniel et al. (4) use

H. Sol and C.W.J. Oomens (eds.), Material Identification Using Mixed Numerical Experimental Methods, 11-20.

the optical microscope to study the failure process of unidirectional composites under transverse and longitudinal loading. Using such techniques a qualitative understanding of microphenomena can be obtained, but for quantitative measurements of microscopic strains Scanning Electron Microscopy (SEM) is a more suitable technique, because of its high resolution and a large depth of field. In situ observation of fracture and damage propagation in ceramic matrix composites using a tensile stage in a SEM has been performed by several researchers (5; 6). A SEM also offers the opportunity to use the electron beam to produce line or dot patterns on the surface of specimens. The deformation of these patterns during loading of the specimen can be used as a measure for the deformation of the material (7; 8; 9).

In this study, an experimental technique has been developed for the measurement of local deformations in a composite. This technique offers the possibility to measure displacement- and strain fields, typically in an area of 20×20 μm, on the surface of a transversely loaded composite specimen. The experimental technique is part of a mixed experimental-numerical method, which is used to characterize the mechanical properties of interphases. This method is described in the next section. Subsequently, the measurement technique is described, after which a number of applications for the characterization of microphenomena are given. Finally some conclusions are offered.

2 A mixed experimental-numerical method

The method is based on the combination of the measurement of displacements, finite element modelling, and parameter estimation. The field quantities are measured during transverse experiments on unidirectional E-glass-fibre-reinforced epoxies. Figure 1 shows a schematic representation of the method.

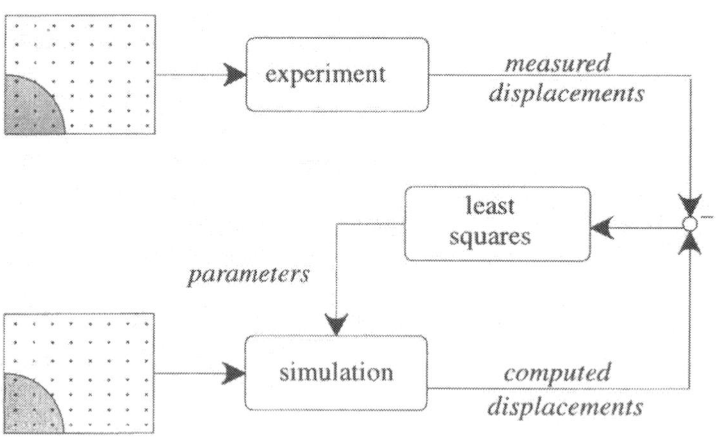

Figure 1. Schematic representation of the mixed experimental-numerical method.

Measurement of field quantities is an essential part of this study and will be discussed in detail in the next section. A grid of markers is placed on the surface of a specimen, and displacements of the markers are measured during tensile experiments in a SEM. A finite element model of the experiment is used to calculate the displacements. The geometry of this model is defined by the positions of the border markers (Figure 2). The displacements of these markers define the kinematic boundary conditions. This method offers the advantage that the numerical model gives an exact description of the local stress-strain state in the experiment. Measured and computed displacement fields are linked in order to find material parameters during an iterative procedure. The difference between measured and computed displacements is used to adjust the parameters describing the constitutive behaviour of the material.

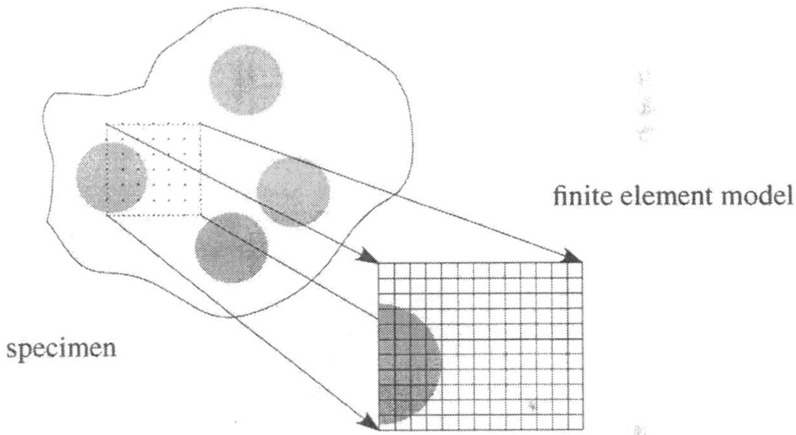

Figure 2. Local modelling of a part of a specimen.

The estimation of interphase parameters comprises the comparison between the experimentally obtained displacement data of the inner markers of the grid (Figure 2) and the results of the finite element analysis (10). This comparison reveals updated estimations of the material parameters in the simulation via a least squares algorithm. Using this estimation procedure, unknown mechanical properties of interphases can be determined, when a mathematical model of the experiment is available.

The numerical simulation has to give an accurate representation of the stress-strain state in the experiment. Therefore orienting simulations were performed to analyze the stress-state in transversely loaded composites, and to test the influence of the marker-field position on the accuracy of the estimated material parameters.

Because displacements are measured at the surface of a transversely loaded specimen, a 3D square finite element model consisting of a quarter of a glass fibre and the surrounding polymer matrix has been used for the analysis of 3D effects. Figure 3 shows the maximum radial stress at the fibre-matrix interface, as a function of the distance to the specimen surface for three different fibre volume fractions.

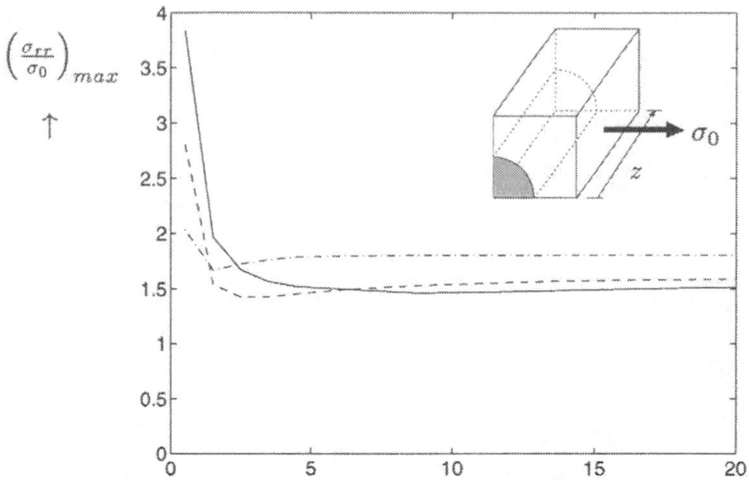

Figure 3. *Maximum radial interfacial stress as a function of the distance to the specimen surface.* $\cdot - \cdot V_f = 45\%, - - - V_f = 20\%, — V_f = 5\%$

The maximum radial stress increases significantly towards the specimen surface (z=0), showing the necessity of using 3D models for these type of experiments. At a distance of 15 μm from the surface, the stress is constant. Furthermore, the stress built-up towards the surface increases for a decreasing fibre volume fraction. When these results are translated to high volume fraction composites with a inhomogeneous fibre distribution, this would imply that the interfacial radial stresses are the highest in the matrix rich regions. Based on these numerical results, all experiments are modelled using 3D finite element models, with a thickness of 15 μm.

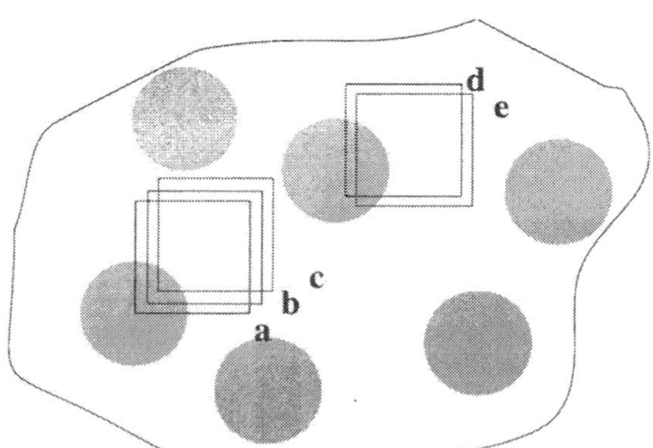

Figure 4. *Numerical simulations of different marker field positions.*

Given a 3D-model of an experiment, numerical simulations have been performed to check the influence of changes in experimental setup and the influence of measurement noise on the estimation of parameters. Several different geometries have been analyzed (Figure 4).

For all geometries (a-e), the markers are placed in a square packing, and marker distances of 1 μm and 2 μm have been used. In all models, the thickness of the interphase was assumed to be 1 μm, and the material properties of glass-fibre (E=70 GPa), interphase, and epoxy matrix (E=2.5 GPa) were modelled linearly elastic. For the interphase, a Young's modulus of 2.0 GPa was taken as the 'real' value. With these values, the in-plane displacements of the markers were calculated, with prescribed displacements on the boundaries of the models. Subsequently, the marker displacement field and the boundary conditions were disturbed by random noise with an amplitude of 5% or 10% of the maximum displacement value. The disturbed displacements were used to estimate the Young's modulus of the interphase. For the initial estimate, a value of $\hat{E}_0$=1.0 GPa was chosen, and for each experimental setup 10 different random noise realizations were added to the calculated displacements. For the setups with a measurement noise of 5%, the average values of $\overline{\hat{E}_i}$ are close to the 'true' value of 2 GPa. Even for a measurement noise of 10% the maximum error was 0.2 GPa. The standard deviation of the 10 different estimates increases from approximately 0.3 GPa to 0.7 GPa for the increasing amplitude of the measurement noise.

3 A measurement techinique for the micronscale

A procedure has been developed to conduct near real-time measurements of deformations of composite materials, and to subsequently obtain measurements of displacements on the micronscale. A grid of markers on the surface of test specimens can be generated in the SEM, as is described in this section. For analysis in the SEM, polished composite specimens were used. To avoid charging, a ±100 Å thick electrically conductive gold-palladium coating was applied. Two types of markers can be obtained: raised dots and black spots.

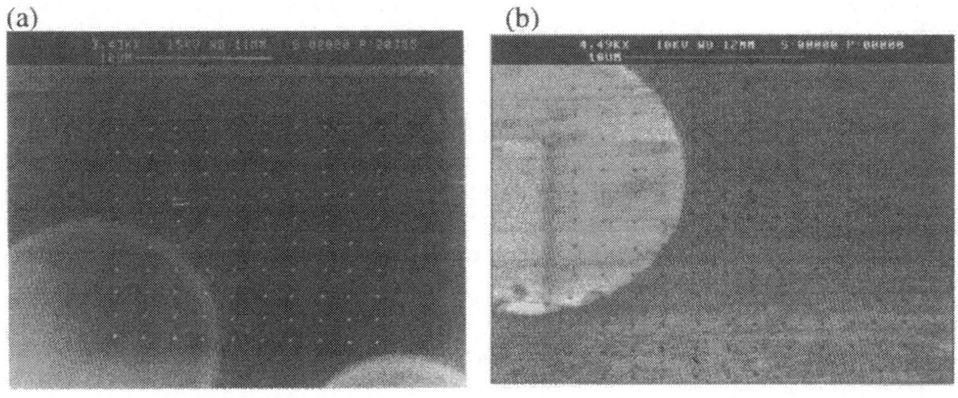

Figure 5. Two types of markers on a E-glass/epoxy composite: (a) raised dots, (b) black spots.

During the observation of a specimen at high magnifications (>5000X), the surface of the specimen becomes coated with a dark layer of contamination present in the vacuum chamber of the SEM. This contamination has a relatively high coefficient of thermal expansion, which can be exploited when the SEM is used in spot mode. The contamination is locally heated by the electron beam and expands, resulting in a marker, being a white spot, which represents a local raised dot on the surface. Repeating this procedure produces a marker field on the surface of a composite (Figure 5a). The dots are placed on a part of a glass fibre cross-section and on the epoxy matrix. The diameter of the dots in Figure 5 is 0.2 μm, and they are obtained with an acceleration voltage of 10KV, a magnification of 10000X, and a dwell time of 60 seconds.

The second type of markers were obtained with the same settings, but with a reduced dwell time. The beam is now only used to concentrate an amount of contamination to one spot on the specimen surface. An example of a marker field consisting of these black spots is given in Figure 5b. The dwell time varied for different specimens between 5 and 10 seconds. In the characterization of microphenomena described in this paper, mostly dot maps consisting of these black spots have been used. Displacements are measured by comparing the coordinates of the dots in reference and deformed situation. The coordinates are determined from directly digitized images. After isolating the markers from the images, the coordinates were defined to be their centre of gravity.

4 Characterization of Interphase Conditions

Of all composite properties, especially the properties of the interphase are difficult to determine. Up to now, most of the work to characterize these properties has been done in shear loading conditions, using experiments such as pull-out, microdebond, fragmentation or microindentation tests. However, since transverse debonding is often the first damage mechanism in composite structures, it is important that also the interphasial properties in normal loading conditions are characterized. For this purpose, glass fibre reinforced composite plates were fabricated using the filament winding technique. In order to observe microscopic displacements in-situ during experiments in the SEM, dumpbell shaped miniature specimens with a small gauge length were designed. The edges of the specimens were polished using standard metallographic procedures, to a grit size of 1 μm, after which the specimens were coated with a Au-Pd sputter coating. Dot maps were placed on a part of a fibre and the surrounding matrix material in the gauge section of the specimen. Monotonic tensile tests are carried out in the SEM. The motor of the tensile stage was stopped at increasing load steps. At each load step the coordinates of the markers are determined. An example of an experiment is shown in Figure 6. The displacements shown in Figures 6b and 6c are measured near the central fibre in Figure 6a for global loads of 100 N and 150 N, respectively. The maximum displacements were 0.12 μm and 0.29 μm for the second load step. The finite element model of this experiment is shown in Figure 6d. The geometry is defined by the coordinates of the border markers. The displacements of these

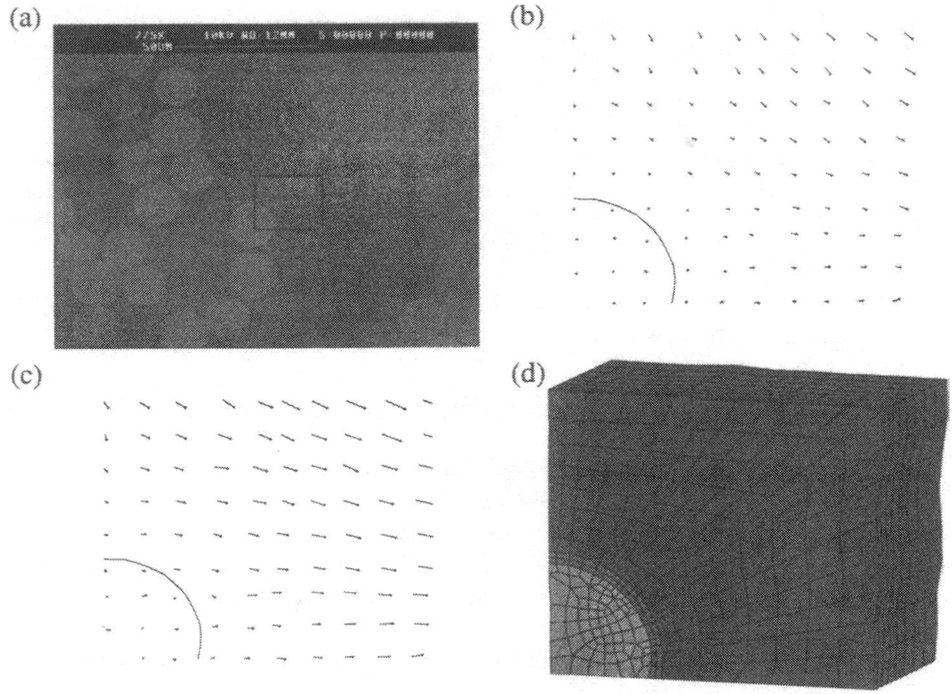

Figure 6. *Experiment near a matrix rich region: (a) overview, (b) measured displacements ×5 for F=100N, (c) measured displacements ×5 for F=150N, (d) finite element model of the experiment.*

markers were used as boundary conditions for the nodes with equal in-plane coordinates, and for the boundary nodes that do not coincide with markers the displacements were interpolated. The displacements of the inner markers on the surface of the model were used to estimate the Young's modulus of the interphase. The interphase was assumed to be isotropic and having a thickness of 1 μm. Table 1 shows the results for this experiment.

TABLE 1. *Estimated interphase properties*

load [N]	max. disp. [μm]	E_f [GPa]	E_m [GPa]	$\bar{\hat{E}}_i$ [GPa]	stdv. [GPa]
100	0.12	70	2.8	0.93	0.48
150	0.29	70	2.8	0.33	0.16

At each load step, 10 digital images of the marker field were stored, from which the coordinates of the markers were determined from image analysis, resulting in 10 different displacement fields with respect to the reference situation. The difference is caused by the measurement noise. For each of the 10 displacement fields E_i was estimated using the experimental-numerical method. In all cases, the estimated values $\hat{E}_i$ converged to a constant value within 5 iterations.

For the second load step, the average of $\hat{E}_i$ is lower than for the first load step. The measured displacements in the second load step lead to small plastic strains in the interphase, which can be observed in the numerical simulations. Fitting with a linear elastic constitutive model then leads to a lower value of $\overline{\hat{E}_i}$. Therefore, the first value of $\overline{\hat{E}_i}$ is closest to the true value of the Young's modulus of the interphase for this composite.

5 Determination of the Interfacial Normal Strength

Transverse loading of unidirectional composites leads to initiation of microcracks at fibre-matrix interfaces, followed by propagation of these cracks leading to macroscopic failure. Therefore, the interfacial normal strength (INS) is an important material parameter. In the experiments presented here, single glass fibre model composites are used. During transverse loading of these specimens in the SEM, initiation of debonding and crack-growth can be observed. The measurement technique presented in this paper was used to measure the local stress-strain state just before crack initiation. Figure 7 shows such a measured displacement field, and the initiation of debonding just after that load step.

(a) (b)

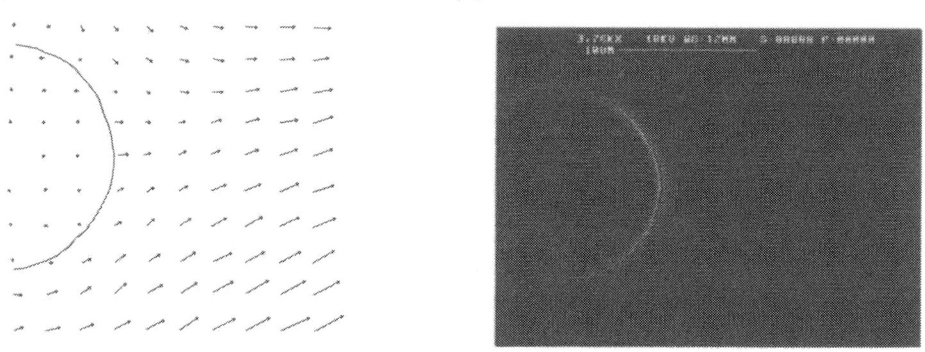

Figure 7. Experiment on a single fibre: (a) displacement field before debonding (×5), (b) initiation of debonding.

The maximum measured displacement in this experiment was approximately 0.4 μm, resulting in plastic deformations around the fibre. In the finite element model of this experiment, ideal plastic behaviour was assumed for the matrix material with a Young's modulus of 2.8 GPa and a Von Mises yield stress of 66 MPa.
The geometry of the model was defined by the coordinates of the border markers, and the displacements of these markers were used as boundary conditions. The

model with (measured) boundary conditions was used to calculate the radial and tangential stresses at the fibre-matrix interface. Because the stresses are the highest at the specimen surface, these stresses were calculated at the surface.

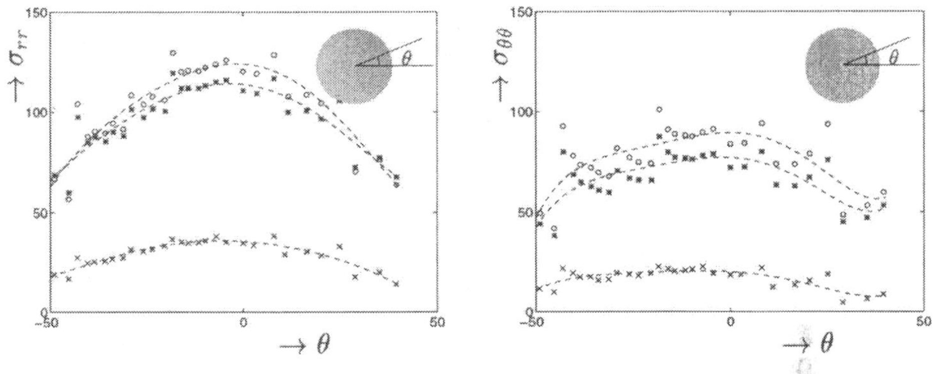

Figure 8. *(a) Radial and (b) tangential interfacial stresses for increasing global load.* $\times\times\times$=*77 N*, $***$=*151 N*, o o o=*198 N*.

The results are given in Figure 8, and show that the radial interfacial stress increases for increasing global loads, and that the radial stress reaches its maximum value of 125 MPa at an angle θ of -5°. Similar stress distributions were found in numerical simulations performed by de Kok et. al. (11). For all load steps the radial stresses are higher than the tangential stresses, which results in initiation of a crack along the fibre surface.

6 Conclusions

In this paper, an experimental-numerical method for the characterization of microphenomena in composite materials has been presented. The method is based on the measurement of field quantities during transverse experiments in a SEM. Displacements were measured on a scale of microns using markers that are produced with the electron beam of the SEM. The dot maps on the surface of specimens define the geometry and boundary conditions of a finite element model of the experiment. The advantage of this approach is that the model gives an exact description of the local stress strain state in the specimen. A number of applications for the method have been presented. Interphase conditions in high fibre volume fraction composites were estimated with an iterative procedure. An important application is the determination of the interfacial normal stress (INS). For this purpose, the measurements were used to define the local stress strain state in single fibre composites just before the moment of initiation of debonding. A finite element model was used to calculate the radial and tangential stresses at the fibre matrix interface. Based on the presented results, it can be concluded that the presented method is a suitable tool for the characterization of various microphenomena in composite materials.

Acknowledgement
This work is part of a research project financially supported by the Dutch Technology Foundation (STW).

REFERENCES

[1] de Kok, J. M. M., Meijer, H. E. H. and Peijs, T., 'The influence of matrix plasticity on the failure strain of transversely loaded composite materials' *Proc. ICCM/9, 1993, pp. 242-249*

[2] Dibenedetto, A. T. and Jones, K. D., 'The Role of Interphase Debonding on Cumulative Fractures in a Continuous Fibre-Reinforced Composite' *Comp.,* **27A**, *1996, pp. 869-879*

[3] Heuvel, P. W. J. van den, Bruggen, Y. J. W. van der and Peijs, T. 'Failure Phenomena in Multi-fibre Model Composites: Part 1. An Experimental Investigation into the Influence of Fibre Spacing and Fibre-Matrix Adhesion' *Comp.,* **27A**, *1996, pp. 855-859*

[4] Daniel, I. M. Anastassopoulos, G. and Lee, J.-W. 'Experimental Micromechanics of Brittle-Matrix Composites' *Micromechanics: experimental techniques, ASME (AMD vol. 102), pp. 133-146*

[5] Dalmaz, A., Reynoud, P., Rouby, D. and Fantozzi, G. 'Damage Propagation in Carbon/Silicon Carbide Composites During Teinsile Tests Under the SEM' *J. Mat. Sci.,* **31**, *1996, pp. 4213-4219*

[6] Shercliff, H. R., Beaumont, P. W. R., Vekinis, G. 'Direct Observation of the Fracture of CAS-glass/SiC Composites, part II Notched Tension' *J. Mat. Sci.,* **29**, *1994, pp. 4184-4190*

[7] Corleto, C. R., Bradley, W. L. and Brinson, H. F. 'An Experimental Micromechanics Measurement Technique for Sub-micron Domains' *J. Mat. Sci.,* **31**, *1996, pp. 1803-1808*

[8] Dally, J. W. and Read, D. T. 'Electron Beam Moiré' *Exp. Mech.,* **33**, *1993, pp. 270-277*

[9] Read, D. T. and Dally, J. W. 'Electron Beam Moiré Study of Fracture of a Glass Fiber Reinforced Plastic Composite' *J. Appl. Mech.,* **61**, *1994, pp. 402-409*

[10] Meurs, P. F. M., Schreurs, P. J. G., Peijs, T. and Meijer, H. E. H. 'Characterization of Interphase Conditions in Composite Materials' *Comp.,* **27A**, *1996, pp. 781-786*

[11] de Kok, J. M. M. van Klinken, E. J. and Peijs, T. 'Influence of fibre surface treatment on the transverse properties of carbon fibre reinforced composites' *Adv. Comp., 1993, pp. 427-432.*

IDENTIFICATION OF SHEAR-MODULI OF MICROSTRUCTURES

G. SCHILTGES, D. GSELL, J. DUAL
Institute of Mechanics
ETH Zürich, Swiss Federal Institute of Technology
CH-8092 Zürich, Switzerland

1 Abstract

In this investigation special care is paid to the shear-moduli of microstructures. The specimens used are either made of monocrystalline silicon, nickel or nickel-iron alloys, with the following dimensions: height: 40 - 200 μm, width: 20 - 90 μm, length: 200 - 1000 μm.

In a special torsional setup these structures are twisted until fracture. Both torque and rotation angle are registered simultaneously. A micromachined device with differential force measurement gives a resolution of 0.3 μNm, whereas a laser-optical system allows twisting angles of 0.03° to be detected. G_{12} and G_{13} can be determined from the experiments if the results of two or more cross sections are combined. Because of the complicated shape of the structures finite element calculations are supplemented with the exact analytical solution for the torsion problem in prismatic orthotropic bars. In an iterative process involving experimental, numerical and analytical features, it is then possible to match theory and experiment efficiently and thus identify G_{12} and G_{13}.

Applied to silicon structures this method yields good results. As expected from its cubic symmetry both moduli have the same value ($G_{12} = G_{13}$, difference < 2%) and are in good agreement with values found in dynamic measurements.

Metallic specimens produced by the LIGA-technique can be shown to have orthotropic behaviour with G_{12} less than $\frac{G_{13}}{3}$.

2 Introduction

Micromechanics is a fast growing field, as more and more components are manufactured at micrometer size. Sensors, actuators, microvalves, micropumps and micromotors are but a few of the possible applications already realized. The combination of mechanical and electrical features at a micrometer level (micro electro mechanical systems known as MEMS) has provided the decisive boost to a now thriving industry.

H. Sol and C.W.J. Oomens (eds.), Material Identification Using Mixed Numerical Experimental Methods, 21-30.

A lot of micromechanical structures, i.e. accelerometers, force sensors, nozzles for ink jet or torsional micromirrors [1], [2], [3] are made of silicon using different etching techniques. These techniques are well known from microelectronic processing. A good overview is given by Petersen [3] who also discusses the possibilities of silicon as a mechanical material.

In the 1980's a new technique was developed at the Kernforschungszentrum Karlsruhe, allowing to realize metallic microstructures with high aspect ratios. This process is known as LIGA process and consists mainly of three steps, i.e. deep X-ray lithography, electrodeposition and moulding [4]. Meanwhile the process can be used to fabricate microstructures from metals, polymers, ceramics or even glasses [5], [6].

Due to this great variety of microstructures and their large field of application [5], [7], it is of vital interest to know the mechanical properties of the materials used. Due to new fabrication techniques, i.e. anisotropic etching or LIGA process, microstructures often present different mechanical properties as compared to structures of the same material on a macro-level. One of the main aspects is the establishment of the elastic constants and failure stresses of these new materials.

Petersen et al. [8] try to determine Young's modulus of silicon dioxide by measuring resonance frequencies of microcantilevers. The same approach is used by Putty et al. [9] by using polysilicon microbridges as resonant structures. Similar tests were performed by Hök et al. [10] on various materials. The moduli of the microstructures matched those of the corresponding macromaterials. Static bending tests were used by Weihs et al. [11], [12] and by Tai et al. [13] to measure Young's modulus of different materials.

The advantage of static tests, besides the determination of elastic moduli, is the possibility to establish yield or fracture criteria. This was done by Weihs et al. [11], [12] on silicon, silicon dioxide and gold structures. Koskinen et al. [14] performed tensile tests on polysilicon fibers and gave values for tensile strength and for Young's modulus. Mazza et al. [15] were able to establish yield stress and Young's modulus for LIGA structures by realizing tensile tests and resonance experiments. A good overview of different dynamic and static techniques is given by Schweitz [16].

In this presentation a new static torsional test is presented. The experimental setup is explained and first test results on silicon and LIGA microstructures are shown. By performing experiments on specimens with different cross sectional dimensions, an analysis of the shear moduli is possible. This procedure, involving numerical and analytical data is explained in some detail. First results obtained with this method on both silicon and LIGA specimens are presented. The shear moduli found for silicon are in good agreement with values known from bulk material. No comparison could be made with the shear moduli found for LIGA material, which shows an anisotropic behaviour.

3 Experiment

3.1 SPECIMENS

Two different kinds of materials are used in the experiments, silicon and metal or metal alloys.

Silicon specimens are made by anisotropic wet etching of (100) silicon wafers [17]. The actual testing region has a length of about 250 μm. The cross sectional area is largely determined by the time the wafers are immersed in the KOH etchant. Specimens with two different rectangular cross sections were produced at our institute. They present a width and height of 90 x 50 μm^2 and 50 x 60 μm^2 respectively (Fig. 1). This small bridge is attached to large plates (5 x 5 x 0.38 mm) in order to be able to handle the microstructure (i.e. clamping).

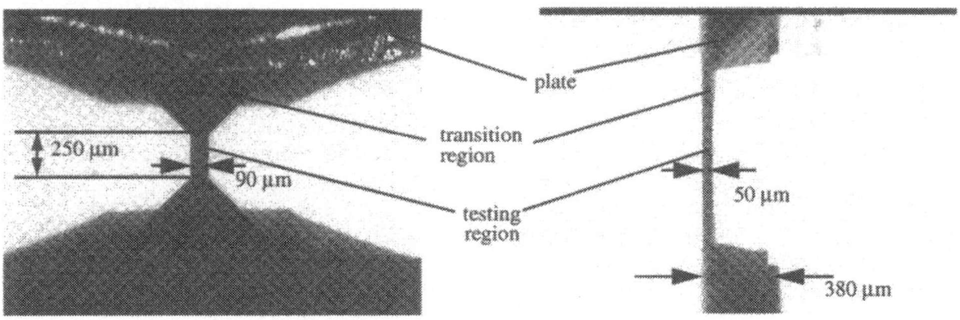

Fig. 1: Silicon specimen

The metal specimens were produced at the IMM in Mainz by means of the LIGA-technology [15]. They have a similar design as their silicon counterparts, with two large plates to assure the handling. The length of the testing region amounts to 300 μm, whereas the width and height may vary between 30 to 40 μm and 70 to 180 μm respectively (Fig. 2). The specimens investigated in this paper are either made of pure Ni or of a NiFe alloy with 30% Fe.

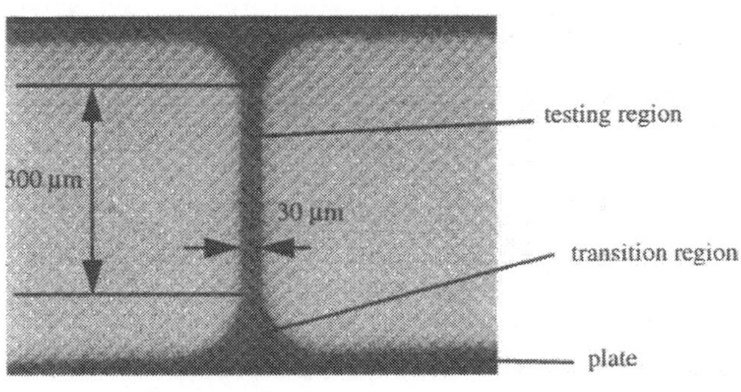

Fig. 2: LIGA specimen

3.2 EXPERIMENTAL SETUP

For the experiments to be conclusive, the setup must be able to assure two main functions, namely apply pure torsion (actuator part) and measure the resulting rotation and torque (sensor part).

The specimen is glued to a clamping interface, which is put on the axis of the torque sensor (see Fig. 3). The sensor used is based on differential force measurement [18]. It has a range of ± 280 μNm with a resolution and accuracy of 0.3 μNm. The sensor is mounted on an x-y translation stage, which is itself fixed on a rotation stage, which has a resolution of $1/1000^{o}$ and a maximum wobble of 70 μrad.

In order to have pure torsion, in particular stresses due to bending need to be minimized. Therefore a minimal wobble is important in order to have nearly parallel axes of the rotation stage and the specimen. Parallel axes are a necessary, but not a sufficient condition to minimize stresses due to bending. Even if both axes are parallel, they may not coincide. Their alignment is obtained by means of the x-y translation stage. The initial position of the specimen axis is recorded with a CCD-camera, which captures a magnified image through a monozoom microscope. In a second step, the specimen, with its lower end still free, is turned by 180° and an image of this second position is taken. By comparing both images with an image processing software (see Fig. 4), the translational error can be reduced to about 2 μm. This procedure has to be repeated twice, with the camera at two different positions, preferably at an angle of 90°.

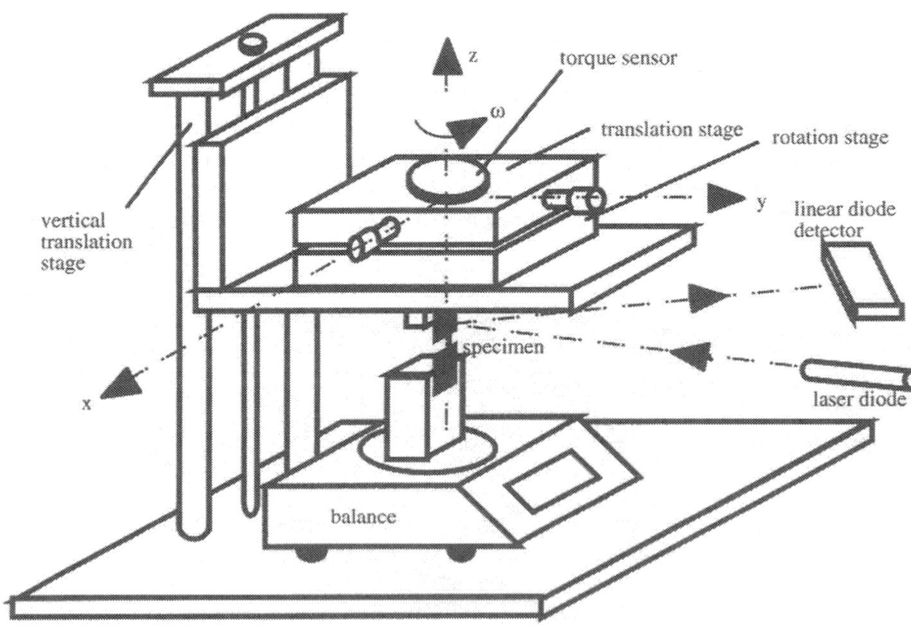

Fig. 3: setup for torsional experiments

After the alignment, the lower part of the specimen is glued to a metal block, which stands on a balance with a resolution of 10^{-5} N. While the glue dries and hardens, it induces tensile stresses in the specimen. The tensile force can be measured with the balance. The specimen is now clamped at both sides and the upper plate can be rotated, while the lower plate remains in its initial position.

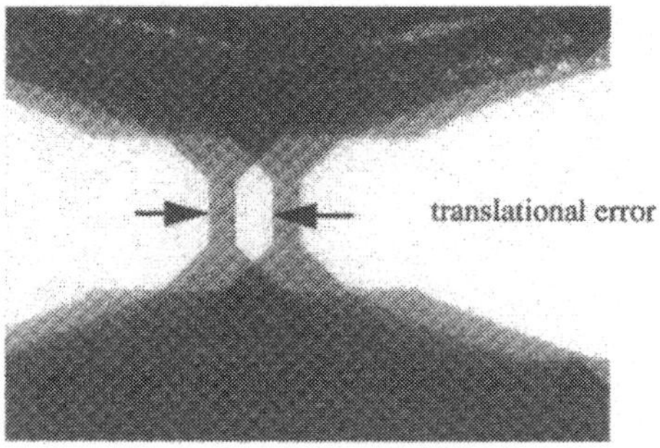

Fig. 4: two superimposed images in the process of alignment

The actual rotation angle is measured with a HeNe-laser and a linear diode detector. The laser beam is adjusted so that its reflection on the specimen surface can be captured with the linear detector. Depending on the rotation of the specimen the beam hits the diode at a different position. The detector has a resolution of 50 µm at a total range of 70 mm. The distance between specimen and detector being about 100 mm, the angular resolution is about 0.03°. This distance changes with each experiment and is hard to determine, so the angular measurement system has to be calibrated before each experiment. This is done while the lower plate is still free. The laser beam is directed on the upper plate and the angular displacement of the rotation stage is taken as a reference to calibrate the signal of the detector. When the specimen is clamped at both ends, it can be shown that the lower end stays rigid and does not rotate. Thus one angular detector is sufficient to know the displacement state of the whole specimen.

3.3 EXPERIMENTAL RESULTS

Due to the gluing of the lower part, a compressive force of about 100 mN acts on all the specimens. In this study only elastic deformations are considered, thus the superimposed tensile stress has no effect on the determination of the shear-moduli. Nevertheless the specimens are rotated till failure occurs, at a speed of 0.025 $^{\circ}$/s. Every two seconds the momentary rotation angle and the corresponding torque are measured. This leads to diagrams like those shown in Fig. 5. As expected silicon fails in a brittle fashion, whereas Ni has an elastic-plastic regime. For both materials the rigidity of the specimen in the elastic case can be calculated from the recorded values.

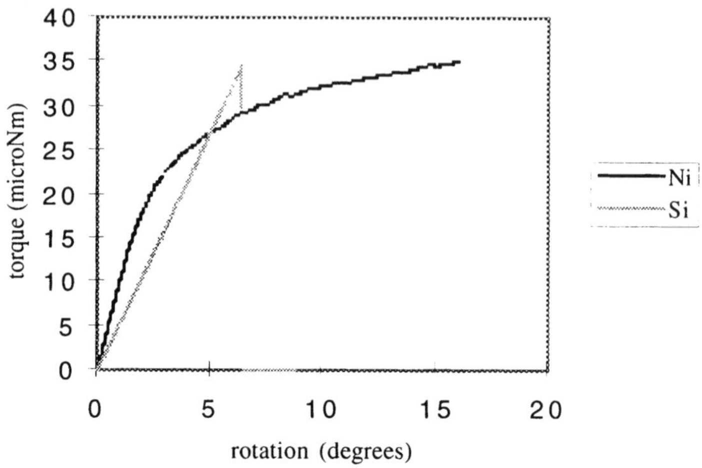

Fig. 5: torque/rotation diagrams for silicon and Ni specimens (only a small part of the diagram for Ni is represented)

4 Determination of shear-moduli of anisotropic material

4.1 TORSION OF AN ANISOTROPIC ROD WITH CONSTANT RECTANGULAR CROSS SECTION

The torsional problem of an anisotropic rod with constant rectangular cross section is described by the generalized Poisson-equation (where $\Phi_{,i} = \frac{\partial \Phi}{\partial x_i}$):

$$\frac{\Phi_{,22}}{G_{13}} + \frac{\Phi_{,33}}{G_{12}} = -2 \tag{1}$$

Φ is the stress function and G_{12} and G_{13} are the governing shear moduli. The axis x_1 is parallel to the axis of the rod, x_3 is perpendicular to the plate and x_2 is perpendicular to x_1 and x_3 (see Fig. 6). For a given rotation angle θ, the corresponding torque T is a function of θ and of Φ:

$$T = 2\theta_{,1} \iint \Phi(G_{12}, G_{13}) dx_2 dx_3 \tag{2}$$

For a rectangular cross-section, the stress function can be computed and introduced in Eq. (2). The torque can then be expressed as an infinite sum (Eq. (3)), which converges rapidly. In Eq. (3) a and b are the half-width and half-height of the rod respectively.

$$T = \frac{2^{12}}{\pi^6}\theta_{,1}ab \sum_{m=1,3}^{\infty}\sum_{n=1,3}^{\infty}\frac{1}{m^2n^2}\left(\frac{m^2}{a^2G_{13}}+\frac{n^2}{b^2G_{12}}\right)^{-1} \tag{3}$$

G_{12} and G_{13} are the two unknown values to be determined. The torque and the rotation can directly be deduced from the experiment. At least two experiments have to be carried out to obtain two equations. If the ratio width/height of the two specimens differs, a system of two linear independent equations results (Eq. (4)), which can then be solved numerically for G_{12} and G_{13}.

$$\begin{aligned} T_1 &= \frac{2^{12}}{\pi^6}\theta_{1,1}a_1b_1 \sum_{m=1,3}^{\infty}\sum_{n=1,3}^{\infty}\frac{1}{m^2n^2}\left(\frac{m^2}{a_1^2G_{13}}+\frac{n^2}{b_1^2G_{12}}\right)^{-1} \\ T_2 &= \frac{2^{12}}{\pi^6}\theta_{2,1}a_2b_2 \sum_{m=1,3}^{\infty}\sum_{n=1,3}^{\infty}\frac{1}{m^2n^2}\left(\frac{m^2}{a_2^2G_{13}}+\frac{n^2}{b_2^2G_{12}}\right)^{-1} \end{aligned} \tag{4}$$

4.2 TORSION OF AN ANISOTROPIC ROD WITH CHANGING RECTANGULAR CROSS SECTION

The system of equations in Eq. (4) is valid only if $\theta_{,1}$ is constant, which means that the cross section of the beam is of constant shape and area. As can be seen in Fig. 1 and Fig. 2 the cross section of the analyzed specimens is not constant. Its width increases at the ends of the microbridge, whereas its height stays the same. Thus Eq. (4) cannot be directly used to specify the shear moduli. The problem can however be solved in a mixed analytical, numerical and experimental approach, whose individual steps will now be explained.

- Two experiments on specimens with a different width/height ratio are carried out and the ensuing torque-rotation relation is recorded.
- A three dimensional finite element model of both specimens is generated. In a first step, as the elastic parameters are unknown, their values have to be guessed. The lower plate is clamped and fixed, and the upper plate is rotated by an angle θ_u. The value of $\theta_{,1}$ changes over the transition region but remains constant, $\theta_{r,1}$, in the testing region. The value of $\theta_{r,1}$ is obtained by taking the rotation of two cross sections and by dividing the difference of rotation by the length l between the sections (see Fig. 6):

$$\theta_{r,1} = \frac{\theta_{S2}-\theta_{S1}}{l} \tag{5}$$

- Eq. (4) can now be used. The values of $\theta_{1,1}$ and $\theta_{2,1}$ are known from the FE simulation. The torques T_1 and T_2 can be extracted from the experimental data. They correspond to the angles θ_{u1} and θ_{u2} by which the upper plates of the FE models are rotated. By resolving Eq. (4) with theses values, a first approximation of the shear moduli is obtained.

- The calculated shear moduli are introduced in the FE simulation, assuming homogeneous properties for the whole specimen, and a new value of $\theta_{r,1}$ is computed for both specimens.
- The two previous steps are repeated till the difference between the shear moduli of two consecutive iterative processes is small enough.

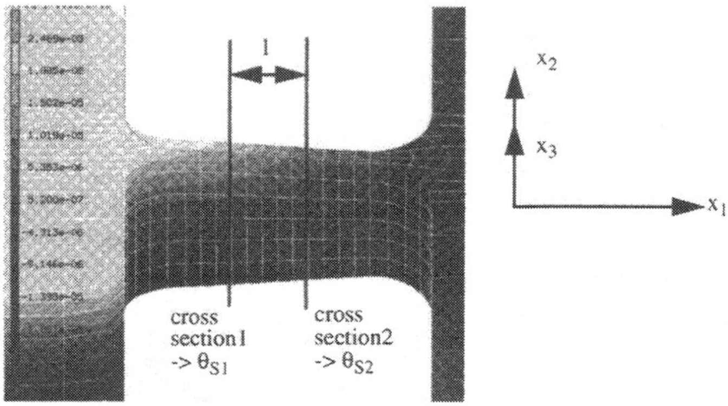

Fig. 6: FE-image of a LIGA specimen under torsion (grey tones show displacement in x_2 direction)

4.3 APPLICATION TO SILICON AND LIGA MICROSTRUCTURES

The iterative process described in section 4.2 is first tested on monocrystalline silicon specimens. One specimen has a cross section of 50 x 59 μm^2, while the other's dimensions are 88 x 48 μm^2. These values were measured with a UBM-profilometer, based on a laser focussing principle, with a resolution of about 1 μm. When beginning with a start value for the shear moduli of 65 GPa, the iteration converges already after three steps to G_{12} = 75 GPa and G_{13} = 74 GPa. These values are confirmed by tests with other specimens.

Compared to the shear moduli of the bulk material, $G_{12} = G_{13}$ = 79.6 GPa, the measured moduli are always 6 or 7% too small. This may be due to the problem that it is very hard to measure the dimensions of the samples with high accuracy. It can nevertheless be said, that the procedure gives satisfactory results.

After the tests with silicon, the shear moduli of LIGA specimens are specified. The materials of interest are pure Ni and a NiFe alloy with 30% Fe. For Ni the following values are found: G_{12} = 70 GPa, G_{13} = 66 GPa. This is in good agreement with values known from isotropic bulk material, where G amounts to G = 70 GPa. The alloy shows completely different results, G_{12} = 15 GPa and G_{13} = 54 GPa, which is an accentuated anisotropic behaviour.

5 Conclusion

An experimental setup for torsional experiments on microstructures has been presented. Its key features are the possibility of alignment of rotation axis and specimen axis, as well as the separate measurement of both torque and rotation angle with high resolution.

The recorded data allows the determination of the governing shear moduli in a mixed analytical, numerical and experimental procedure. Under the condition that the experiment is run on two specimens with different width/height ratios, a system of two linear independent equations can be established. With data recorded during the experiment and with data from finite element simulations, the system is solved for the shear moduli G_{12} and G_{13}.

To test the setup as well as the iterative procedure, first tests have been run with silicon specimens. Good agreement between the shear moduli determined in this investigation and the known shear modulus of monocrystalline silicon is reached. Metallic specimens made with the LIGA technique are also tested. Though pure Ni shows an isotropic behaviour with a shear modulus similar to that of bulk material, NiFe (Fe 30%) has a strong anisotropic behaviour.

6 Reference

[1] Jaecklin V.P., Linder C., de Rooij N.F., Moret J.M., Vuilleumier R. (1994) Line-addressable torsional micromirrors for light modulator arrays, *Sensors and Actuators A* **41-42**, 324-329

[2] Fischer M., Graef H., von Münch W. (1994) Electrostatically deflectable polysilicon torsional mirrors, *Sensors and Actuators A* **44**, 83-89

[3] Petersen K. (1982) Silicon as a Mechanical Material, *Proceedings of the IEEE* **70**(5), 420-453

[4] Becker E.W., Ehrfeld W., Hagmann P., Maner A., Münchmeyer D. (1986) Fabrication of microstructures with high aspect ratios and great structural heights by synchrotron radiation lithography, galvanoforming, and plastic moulding (LIGA process), *Microelectronic Engineering* **4**, 35-56

[5] Ehrfeld W., Lehr H. (1995) Deep X-Ray Lithography for the Production of three-dimensional Microstructures from Metals, Polymers and Ceramics, *Radiation Physics and Chemistry* **45**(3), 349-365

[6] Ehrfeld W., Abraham M., Ehrfeld U., Lacher m., Lehr H. (1994) Materials for LIGA Products, *IEEE micro electro mechanical systems: an investigation of micro structures, sensors, actuators, machines and robotic systems: proceedings: Oiso, Japan, January 25-28, 1994*, 86-90

[7] Bacher W., Menz W., Mohr J. (1995) The LIGA Technique and Its Potential for Microsystems-A Survey, *IEEE: transactions on industrial electronics* **42**(5), 431-441

[8] Petersen K.E., Guarnieri C.R. (1979) Young's modulus measurements of thin films using micromechanics, *J. Appl. Phys.* **50**(11), 6761-6766

[9] Putty M.W., Chang S.C., Howe R.T., Robinson A.L., Wise K.D. (1989) One-Port Active Polysilicon Resonant Microstructures, *IEEE micro electro mechanical systems, Salt Lake City, Utah, February 20-22*, 60-65

[10] Hök K., Gustaffson K. (1985) Vibration Analysis of Micromechanical Elements, *Sensors and Actuators* **8**, 235-243

[11] Weihs T.P., Hong S., Bravman J.C., Nix W.D. (1988) Mechanical deflection of cantilever microbeams: A new technique for testing the mechanical properties of thin films, *J. Mater. Res* **3**(5), 931-942

[12] Weihs T.P., Hong S., Bravman J.C., Nix W.D. (1989) Measuring the strength and stiffness of thin film materials by mechanically deflecting cantilever microbeams, *Thin Films: Stresses and Mechanical Properties (Mat. Res. Soc. Symp. Proc.* **130**, Pittsburgh, PA, 1989), 87-92

[13] Tai Y., Muller R.S. (1990) Measurement of Young's modulus on microfabricated structures using a surface profiler, *IEEE micro electro mechanical systems, Napa Valley, California, February 11-14*, 147-152

[14] Koskinen J., Steinwall J. E., Soave R., Johnson H.H. (1993) Microtensile testing of free-standing polysilicon fibers of various grain sizes, *J. Micromech. Microeng.* **3**,13-17

[15] Mazza E., Abel S., Dual J. (1996) Experimental determination of mechanical properties of Ni and Ni-Fe microbars, *Microsystem Technologies* **2**, 197-202

[16] Schweitz J. (1992) Mechanical Characterization of Thin Films by Micromechanical Techniques, *MRS Bulletin*, July

[17] Mazza E., Danuser G., Dual J. (1996) Light optical deformation measurements in microbars with nanometer resolution, *Microsystem Technologies* **2**, 83-91

[18] Gass V., van der Schoot B.H., Jeaneret S., de Rooij N.F. (1994) Micro-torque sensor based on differential force measurement, *IEEE micro electro mechanical systems, Oiso, Japan, January 25-28*, 241-244

IDENTIFICATION OF THE DYNAMIC MATERIAL PROPERTIES OF COMPOSITE SANDWICH PANELS WITH A MIXED NUMERICAL / EXPERIMENTAL TECHNIQUE

I. PEETERS, H. SOL
Free University Brussels (VUB), Faculty of Applied Sciences
Pleinlaan 2, 1050 Brussels

1. Introduction

Since the eighties a mixed numerical/experimental technique (MNET) for the determination of the complex moduli of thin plates (Love-Kirchoff hypothesis) is developed at the department of Structural Analysis of the Free University Brussels (VUB) ([1],[2],[3]). It is a user-friendly and fast measuring technique based on the measurement of resonant frequencies and modal damping ratios and is called 'the Resonalyser procedure'. This innovative test method determines the elastic constants and the damping characteristics of a free vibrating plate. The results obtained for the material data can then be used as input data for the analysis (e. g. in a finite element model) of more complex structures made out of the tested plate material.
The aim of the recent work is to develop a similar measuring technique for the estimation of the material characteristics of sandwich panels. Therefore a suitable model that can calculate the needed material data starting from a good chosen value of the parameters of that model has to be established. In this paper the latest work done to obtain an appropriate numerical model that can describe the dynamic behavior of a sandwich in an accurate but relatively simple way is presented. The model has to account for shear deformation because the effect is now significant and can not be neglected as in thin plate theory. In 1993 Hua H. started with the development of a sandwich model [4]. This model that is equivalent with the Reissner-Mindlin thick plate theory is further investigated.
The sandwich model is used for the dynamic analysis of well-known self made panels. At first instance sandwiches with an isotropic core and ditto faces are treated. In the near future sandwiches with orthotropic and anisotropic composite faces will be studied. The frequencies obtained with the numerical model are compared with the experimental ones. The next work to be done is calculating the modal damping ratios of the considered eigenmodes by making use of the energy method ([6],[7]). The work strategy is already explained is this paper. Again, a comparison between the calculated and the experimental determined quantities must be carried out. The comparison

H. Sol and C.W.J. Oomens (eds.), Material Identification Using Mixed Numerical Experimental Methods, 31-39.

between calculations and experiments is necessary to see whether the considered sandwich model is suitable to be used in a MNET for sandwich panels.

2. Numerical Model

For the further understanding of the sandwich theory, figure 1 gives an image of the considered sandwich and the situation of the coordinates. In the formulas, index c is used for the core characteristics and index f for the properties of the faces.

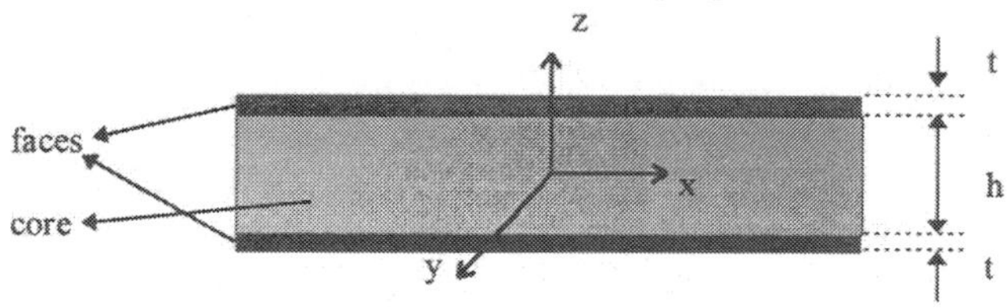

Figure 1. View of the considered sandwich and the coordinate system

2.1 EXPRESSION OF THE DISPLACEMENT COMPONENTS

The investigated theory is based on the following assumptions :

1. The material properties of the faces and the core are symmetric to the $z = 0$ plane;
2. As the faces are very thin, σ_x, σ_y and τ_{xy} are uniformly distributed along the thickness direction;
3. In the core $\sigma_x = \sigma_y = \tau_{xy} = 0$ and the core is only subjected to transverse shear forces;
4. Under flexural deformation (anti-symmetric with respect to the mid-surface) $\varepsilon_z = 0$ in both faces and core;
5. The stress σ_z can be neglected.

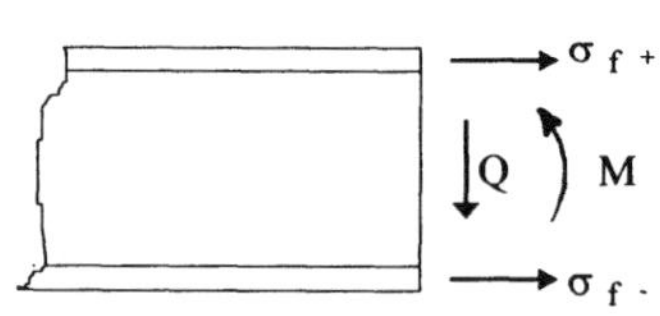

Figure 2. Moment and force distribution and sign convention

With assumption 1, the coupling of in-plane and bending deformation no longer exists. In fact, the in-plane deformation modes are not important in the frequency domain. Assumption 3 together with the equilibrium equation gives for the core :

$$\frac{\partial \tau_{xz}^{c}}{\partial z} = \frac{\partial \tau_{yz}^{c}}{\partial z} = 0 \tag{1}$$

So τ_{xz}^c and τ_{yz}^c are uniformly distributed along the thickness direction. The shear stresses should be (see [5]) :

$$\tau_{xz}^c = G_{xz}^c \gamma_{xz}^c = \frac{h+t}{h} G_{xz}^c \gamma_{xz} \tag{2a}$$

$$\tau_{yz}^c = G_{yz}^c \gamma_{yz}^c = \frac{h+t}{h} G_{yz}^c \gamma_{yz} \tag{2b}$$

The shear forces are given by :

$$Q_x = \tau_{xz}^c h \quad \text{and} \quad Q_y = \tau_{yz}^c h \tag{3}$$

Combining (2) and (3) gives :

$$\gamma_{xz} = \frac{Q_x}{G_{xz}^c (h+t)} \quad \text{and} \quad \gamma_{yz} = \frac{Q_y}{G_{yz}^c (h+t)} \tag{4}$$

The shear strains are also defined by the following expressions (u, v and w are the displacements in the x, y and z direction respectively).

$$\gamma_{xz} = \frac{\partial w}{\partial x} + \frac{\partial u}{\partial z} \quad \text{and} \quad \gamma_{yz} = \frac{\partial w}{\partial y} + \frac{\partial v}{\partial z} \tag{5}$$

As the mid-surface displacements u_o , v_o are zero (pure bending), the next expressions for the displacements in the x and y direction can be found after integration :

$$u = -z\left(\frac{\partial w}{\partial x} - \frac{Q_x}{G_{xz}^c (h+t)}\right) \tag{6a}$$

$$v = -z\left(\frac{\partial w}{\partial y} - \frac{Q_y}{G_{yz}^c (h+t)}\right) \tag{6b}$$

Setting

$$\theta_x = \left(\frac{\partial w}{\partial x} - \frac{Q_x}{G_{xz}^c (h+t)}\right) \quad \text{and} \quad \theta_y = \left(\frac{\partial w}{\partial y} - \frac{Q_y}{G_{yz}^c (h+t)}\right) \tag{7}$$

the displacements can be written as :

$$\begin{aligned} u &= -z\theta_x \\ v &= -z\theta_y \\ w &= w(x,y) \end{aligned} \tag{8}$$

With assumption 4 and equation (8) it can be seen that θ_x and θ_y are independent of z. From equations (4), (7) and (8) follows that a straight line normal to the middle plane of the core before deformation remains straight after deformation, but has a rotation angle θ_x and θ_y. This is an identical assumption as in the Reissner-Mindlin theory.

2.2 EXPRESSIONS FOR MOMENTS AND SHEAR FORCES

For the upper face (z>0) the displacements in the x and the y direction are given by equation (9).

$$u^+ = -\frac{h+t}{2}\theta_x \text{ and } v^+ = -\frac{h+t}{2}\theta_y \tag{9}$$

The expressions of u^- and v^- (lower face, z<0) are the same, but with an opposite sign. The stresses in the faces for the anisotropic case and for small displacements can be written as follows :

$$\begin{Bmatrix} \sigma_x \\ \sigma_y \\ \tau_{xy} \end{Bmatrix} = \begin{bmatrix} c_{11} & c_{12} & c_{16} \\ c_{12} & c_{22} & c_{26} \\ c_{16} & c_{26} & c_{66} \end{bmatrix} \begin{Bmatrix} \varepsilon_x \\ \varepsilon_y \\ \gamma_{xy} \end{Bmatrix} = \begin{bmatrix} c_{11} & c_{12} & c_{16} \\ c_{12} & c_{22} & c_{26} \\ c_{16} & c_{26} & c_{66} \end{bmatrix} \begin{Bmatrix} \partial u / \partial x \\ \partial v / \partial y \\ \partial u / \partial y + \partial v / \partial x \end{Bmatrix} \tag{10}$$

The c_{ij} with i,j = 1, 2, 6 are the plane stress reduced stiffness components of the anisotropic material. By substituting (9) into (10) the stresses in the upper face are found. The stresses in the lower face can be calculated in a similar way.

The moments are given by :

$$\begin{aligned} M_x &= \frac{1}{2}(h+t)t\left(\sigma_x^+ - \sigma_x^-\right) \\ M_y &= \frac{1}{2}(h+t)t\left(\sigma_y^+ - \sigma_y^-\right) \\ M_{xy} &= \frac{1}{2}(h+t)t\left(\tau_{xy}^+ - \tau_{xy}^-\right) \end{aligned} \tag{11}$$

Substitution of (9) and (10) into (11) gives :

$$\{M\} = \begin{Bmatrix} M_x \\ M_y \\ M_{xy} \end{Bmatrix} = \begin{bmatrix} D_{11} & D_{12} & D_{16} \\ D_{12} & D_{22} & D_{26} \\ D_{16} & D_{26} & D_{66} \end{bmatrix} \begin{Bmatrix} -\partial\theta_x / \partial x \\ -\partial\theta_y / \partial y \\ -(\partial\theta_x / \partial y + \partial\theta_y / \partial x) \end{Bmatrix} = \left[D^f\right]\{\kappa\} \tag{12}$$

with

$$D_{ij} = \frac{c_{ij}(h+t)^2 t}{2} \qquad i, j = 1, 2, 6 \tag{13}$$

and $\{\kappa\}$: vector with plate curvatures

$\left[D^f\right]$: matrix with bending stiffnesses determined by the faces

The expressions for the shear forces can easily be found from equations (4) :

$$\{Q\} = \begin{Bmatrix} Q_x \\ Q_y \end{Bmatrix} = \begin{bmatrix} D_{44} & 0 \\ 0 & D_{55} \end{bmatrix} \begin{Bmatrix} \gamma_{xz} \\ \gamma_{yz} \end{Bmatrix} = \left[D^c\right]\{\gamma\} \tag{14}$$

with

$$D_{44} = G_{xz}^c(h+t) \quad \text{and} \quad D_{55} = G_{yz}^c(h+t) \tag{15}$$

and $\{\gamma\}$: vector with shear strains

$\left[D^c\right]$: matrix with shear stiffnesses determined by the core

Equations (13) and (15) are important to keep in mind as they give the expressions for the parameters of the sandwich model.

2.3 ENERGY FUNCTIONAL, DIFFERENTIAL EQUATIONS AND BOUNDARY CONDITIONS

The derivation of the energy functional is given in reference [5]. Assuming the time-displacement separable solution

$$w(x,y,t) = w(x,y)e^{i\omega t}, \ \theta_x(x,y,t) = \theta_x(x,y)e^{i\omega t}, \ \theta_y(x,y,t) = \theta_y(x,y)e^{i\omega t}$$

it can be written as follows :

$$\pi = \frac{1}{2}\int_A \left({}^\tau\{M\}\{\kappa\} + {}^\tau\{Q\}\{\gamma\} \right) dA - \frac{\omega^2}{2}\int_A \left[pw^2 + I\left(\theta_x^2 + \theta_y^2\right) \right] dA \tag{16}$$

with $(p, I) = \int_{-h/2}^{+h/2}\left(1, z^2\right)\rho_c dz + 2\int_{h/2}^{t+h/2}\left(1, z^2\right)\rho_f dz$

in which ρ_f and ρ_c are the mass densities of the faces and the core separately. 'I' stands for the moment of inertia. The energy functional is thus the same expression as in the Reissner-Mindlin theory. The corresponding governing differential equations and associated boundary conditions could be obtained in the same way as for a Reissner-Mindlin plate by writing the forces and moments equilibrium. They have the same expressions, but different plate rigidity definitions. There are three differential equations, with three unknowns, namely θ_x, θ_y and w.

2.4 NUMERICAL SOLUTION FOR PLATE VIBRATIONS

For the identification of the material properties of rectangular thin plates using free vibrating data, Sol [1] used one finite element with high order Lagrange polynomials as shape functions (to avoid C^1 continuity requirement between elements). The method was proved to be useful in the calculation of resonance frequencies and associated mode shapes for rectangular thin plates. Hua [4] extended this method to sandwich panels.

The Lagrange shape functions $N_i(x,y)$ are to be used as trial functions. By the conventional displacement-based finite element method, the discretization procedure starts from the interpolations of the displacement field variables w, θ_x and θ_y by its nodal values (index i).

$$\begin{Bmatrix} w \\ \theta_x \\ \theta_y \end{Bmatrix} = \sum_{i=1}^{n} \begin{bmatrix} N_i & 0 & 0 \\ 0 & N_i & 0 \\ 0 & 0 & N_i \end{bmatrix} \begin{Bmatrix} w_i \\ \theta_{xi} \\ \theta_{xi} \end{Bmatrix} = \sum_{i=1}^{n} [N_i]\{d_i\} = [N]\{d\} \tag{17}$$

By minimizing the energy functional π (formula 16) with respect to the nodal values the eigenvalue equations are obtained.

$$([K] - \lambda[M])\{d\} = 0 \quad \text{with } \lambda = \omega^2 \tag{18}$$

3. The energy method

The prediction of the modal damping ratios will be carried out with the energy method explained by Ungar and Kerwin[6]. They showed that the modal damping ratio can be written as a function of two energy quantities : the modal deformation energy P_j and the modal damping energy D_j. Expression (19) gives the exact relation.

$$\xi_j = \frac{D_j}{4\pi P_j} \tag{19}$$

The two energy quantities are defined as follows [3] :

$$P_j = \frac{1}{2}{}^{\tau}\{W\}_j\left[K'\right]\{W\}_j \tag{20}$$

$$D_j = 2\pi\frac{1}{2}{}^{\tau}\{W\}_j\left[K''\right]\{W\}_j \tag{21}$$

where {W} : real eigenvector obtained from the dynamic analysis of the sandwich (solution of the eigenvalue problem using only the real part of the stiffness matrix)

[K'] : real part of the stiffness matrix

[K'']: imaginary part stiffness matrix

By substituting equations (20) and (21) in (19) a new expression for the damping ratio ξ associated with the eigenmode j can be found :

$$\xi_j = \frac{{}^{\tau}\{W\}_j\left[K'\right]\{W\}_j}{2{}^{\tau}\{W\}_j\left[K''\right]\{W\}_j} \tag{22}$$

In the near future a program will be written to carry out this calculation.

4. Testing of the sandwich model and discussion of the results

The developed Reissner-Mindlin model for a sandwich is verified by the calculation of the first eigenfrequencies f_j of sandwiches with polycarbonate faces and a PUR core. Two panels are studied : one with faces of 1 mm thickness and one with 3 mm thick faces. The total thickness of the panels is approximately 2 cm and the length and width are 30 cm.

The dynamic analysis of the considered testpanels with the developed model gives the needed eigenfrequencies and eigenmodes. The parameters of the model are the bending stiffnesses D_{ij}. These inputdata are calculated from the measured material properties of the faces and the core. The characteristics of the constituent parts of the sandwich are determined separately with the Resonalyser procedure. The model also needs the dimensions of the panel and the mass density of the faces and the core.

The eigenfrequencies and the modal damping ratios are measured experimentally on the rectangular panels with free boundary conditions.

Table 1 and 2 give the results for the first three eigenmodes (torsion, saddle and breathing mode) of each panel.

TABLE 1. Results obtained for sandwich 1
(faces of 1 mm thickness)

	mode 1	mode 2	mode 3
f_j calculated (Hz)	201.0	314.4	374.2
fj measured (Hz)	216.2	337.9	427.0
δf_j (%)	7	7	12
ξ_j measured	0.00533	0.00472	0.00685

TABLE 2. Results obtained for sandwich 2
(faces of 3 mm thickness)

	mode 1	mode 2	mode 3
f_j calculated (Hz)	213.4	335.1	384.1
fj measured (Hz)	253.9	395.2	484.9
δf_j (%)	16	15	21
ξ_j measured	0.00480	0.00474	0.00556

When the calculated (dynamic analysis with Reissner-Mindlin model) and the experimental measured eigenfrequencies of the first three eigenmodes of the panel are compared, the following differences were noticed :
- the experimental eigenfrequencies are remarkably higher than the calculated ones ;
- the difference is greater for the sandwich with thicker faces ;
- the difference tends to be bigger for the higher eigenmodes.

These discrepancies can have different reasons. First of all the testmaterial has to be fabricated correctly and the used material properties of the constituents of the sandwich need to be correct. More experiments are needed to see if the noted differences could be due to this. There is of course the possibility that the sandwich model has to be adapted. Further investigation of the developed theory will be necessary.

Tests were also performed on a sandwich panel with a thin viscoelastic core and thick glass faces. The vibration measurements were performed in an specially designed oven, so the variation of the frequency and the modal damping ratio of different eigenmodes as a function of temperature could be investigated. The same comparative experiment between the experimental vibration measurements and the dynamic analysis with a suited model will be carried out in the near future. At first instance the Reissner-Mindlin sandwich model will be used therefore. It is possible that this model won't be able to characterize this different type of sandwich, so a new theoretical model has to be used. First a study of the existing models for viscoelastic layer damping will be done.

5. Conclusion

A model for the dynamic analysis of a free vibrating sandwich panel was studied. It is similar as the Reissner-Mindlin thick plate theory and gives a relatively simple description of the behavior of a sandwich panel. The validity of the model is tested by calculating the eigenfrequencies and in the near future also the modal damping ratios (by the help of the energy method) of two self made sandwich plates. It is seen that the measured and the calculated quantities does not match very well. Further research and correction of the model has to be done.

6. Acknowledgments

This work has been carried out with the financial support of the Flemish institute for the benefit of the scientific-technological research in the industry (IWT).

7. References

(1) Sol, H. (1986) *Identification of anisotropic plate rigidities using free vibration data*, Ph.D. dissertation, Department of Structural Engineering, Free University Brussels (VUB)

(2) Vantomme, J. (1992) *A parametric study of material damping in glass fiber reinforced thermosetting plastics*, Ph.D. dissertation, Department of Structural Engineering, Free University Brussels (VUB)

(3) De Visscher, J. (1995) *Identification of the complex stiffness matrix of orthotropic materials by a mixed numerical experimental method*, Ph.D. dissertation, Department of Structural Engineering, Free University Brussels (VUB)

(4) Hua, H. (1993) *Identification of Plate Rigidities of Anisotropic Rectangular Plates, Sandwich Panels and Circular Orthotropic Disks Using Vibration Data*, Ph.D. dissertation, Department of Structural Engineering, Free University Brussels (VUB)

(5) Plantema, F. J. (1966) *Sandwich Construction*, John Wiley & Sons, Inc., New York

(6) Ungar, E. E. and Kerwin, E. M. (1962) Loss factors of viscoelastic systems in terms of energy concepts, *The Journal of the Acoustical Society of America* **34**(7), 954-957

(7) Rikards, R. (1993) Finite Element Analysis of Vibration and Damping of Laminated Composites, *Composite Structures* **24**, 193-204

(8) Sol, H. and Hua, H. (1995) *Effect of the shear deformation on the first ten resonant frequencies of Poisson's plates*, Internal Report, Department of Structural Engineering, Free University Brussels (VUB)

MIXED NUMERICAL AND EXPERIMENTAL STIFFNESS IDENTIFICATION METHODS FOR SOFT TEXTILE COMPOSITES

C. SZOSTKIEWICZ, P. MAILLER, P. HAMELIN
Laboratoire Mécanique & Matériaux L2M,
Université Claude Bernard Lyon I, IUT A Génie Civil,
43 Bd du 11 novembre 1918, 69622 Villeurbanne Cedex, France.
Tel: 33.472.692.130 Fax: 33.478.946.906
E-mail: L2M@iuta.univ-Lyon1.fr

1. Introduction

Because they offer good mechanical characteristics for a lesser weight, textile composites are more and more used for tensile roof structures. Moreover, they are waterproof and translucent, they offer reduced construction times and cost reduction by the lesser weight of the supporting structures and many possibilities for the global shape of the structure (new architecture). But these materials present a structural anisotropy which makes the data set more complex in the determination of their behaviour. The safest mechanical characterization requires an experimental determination using suitable tests. In place, these fabrics are submitted to multiaxial stresses. For that reason, but also because of interactions between warp and weft directions, monoaxial tests are no longer suitable for such structures, even if they are still widely used. That is the reason why all along this work we conducted biaxial tests, which are more relevant. Because no method was able to identify these characteristics from experimental biaxial tests, we proposed a mixed numerical and experimental stiffness identification method based on a set of at least three biaxial tests.

2. Analysis Of The Textile Composite

The material we used all along this work is a plain woven polyester fabric coated with PVC (which is mainly meant to protect the fabric from sun rays). Because of its two perpendicular directions, warp (axe 1) and weft (axe 2), the studied fabric can be considered as an othotropic material (see Figure 1). As the thickness is very small, and the fabric is soft, we can consider that the fabric behaves as a membrane. It means that the bending resistance is non-existent.

H. Sol and C.W.J. Oomens (eds.), Material Identification Using Mixed Numerical Experimental Methods, 41-50.

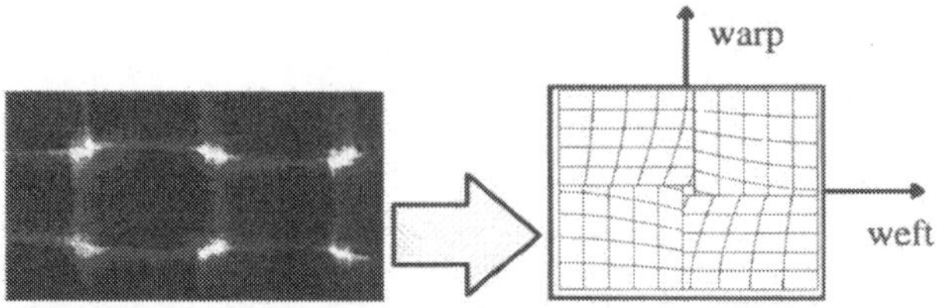

Figure 1. Fabric micro-structure with warp and weft directions.

Assuming an in-plane stress state and a linear stress-strain relation, we apply the thin plate theory to our fabric. We can write the following relation:

$$\begin{bmatrix}\{N\}\\ \{M\}\end{bmatrix}=\begin{bmatrix}[A] & [B]\\ [B] & [D]\end{bmatrix}\cdot\begin{bmatrix}\{\varepsilon\}\\ \{\chi\}\end{bmatrix} \quad \text{with} \quad [A]=\begin{bmatrix}A_{11} & A_{12} & 0\\ A_{12} & A_{22} & 0\\ 0 & 0 & A_{66}\end{bmatrix}$$

where:
$\{N\}$: in-plane stress components
$\{M\}$: moment components
$\{\varepsilon\}$: strain components
$\{\chi\}$: curvatures
$[A]$: in-plane stiffness matrix
$[B]$: in-plane/flexural coupling matrix
$[D]$: flexural stiffness matrix

As the material is symmetrical, the in-plane/flexural coupling matrix [B] are zero. Considering all these assumptions, we will exclusively work on the identification of the in-plane stiffness matrix [A].

Notations: Ewa is the Young's modulus in warp direction (Ewe ... in weft direction) and ν_{12} is the Poisson's ratio between warp and weft.

3. Experimental Set-up

A biaxial tension-tension apparatus was designed and developed by members of L2M Laboratory P. Bonnel, P. Mailler and P. Hamelin, in co-operation with ITF (Institut Textile de France) [1], and thanks to the financial support from Region Rhone-Alpes. The machine (Figure 2) can be controlled either in load or in displacement.

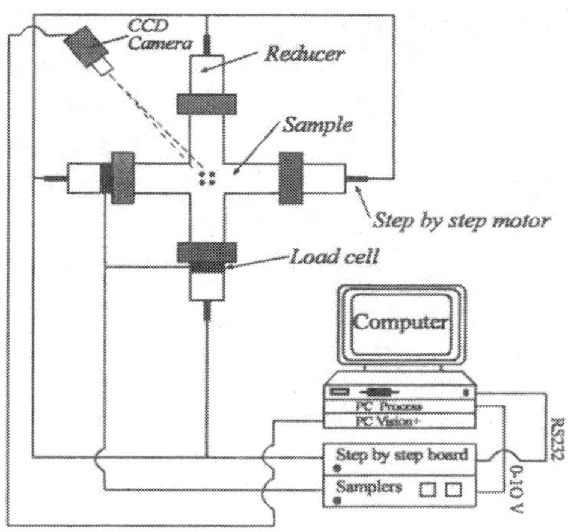

Figure 2. Biaxial apparatus, P. Mailler [2].

Tests are carried out on cruciform (cross-shaped) coated fabric samples, the dimensions of which are given in Figure 3. They are marked with dots in the central part. The large displacements are tracked by an optical measurement of the field with a CCD camera and then processed by a software Visidef (developed by Zhao [3]).

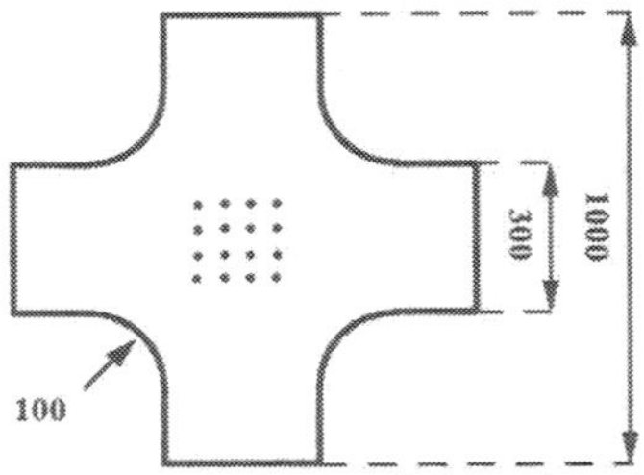

Figure 3. Sample shape and dimensions (mm).

4. Stiffness Identification Method

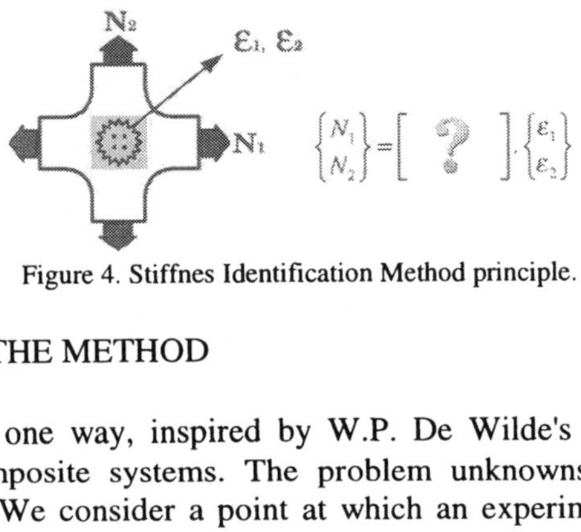

Figure 4. Stiffnes Identification Method principle.

4.1. PRINCIPLE OF THE METHOD

This method was, in one way, inspired by W.P. De Wilde's [4] stiffness properties identification for composite systems. The problem unknowns are Aij, the in-plane stiffness coefficients. We consider a point at which an experimental measurement has been done:

$$\begin{Bmatrix} N_1 \\ N_2 \end{Bmatrix} = \begin{bmatrix} A_{11} & A_{12} \\ A_{12} & A_{22} \end{bmatrix} . \begin{Bmatrix} \varepsilon_1 \\ \varepsilon_2 \end{Bmatrix} \quad \text{in another way:} \quad \begin{Bmatrix} N_1 \\ N_2 \end{Bmatrix} = \begin{bmatrix} \varepsilon_1 & \varepsilon_2 & 0 \\ 0 & \varepsilon_1 & \varepsilon_2 \end{bmatrix} . \begin{Bmatrix} A_{11} \\ A_{12} \\ A_{22} \end{Bmatrix} \qquad (1)$$

Now let us expand the previous relation for n experimental points taken from n different tests:

$$\underset{(2n*1)}{\begin{Bmatrix} N_1^1 \\ N_2^1 \\ \dots \\ \dots \\ N_1^n \\ N_2^n \end{Bmatrix}} = \underset{(2n*3)}{\begin{bmatrix} \varepsilon_1^1 & \varepsilon_1^2 & 0 \\ 0 & \varepsilon_1^1 & \varepsilon_1^2 \\ \dots & \dots & \dots \\ \dots & \dots & \dots \\ \varepsilon_1^n & \varepsilon_2^n & 0 \\ 0 & \varepsilon_1^n & \varepsilon_2^n \end{bmatrix}} . \underset{(3*1)}{\begin{Bmatrix} A_{11} \\ A_{12} \\ A_{22} \end{Bmatrix}} \qquad (2)$$

More simply:

$$\{N\} = [E].\{A\}$$
$${}^t[E].\{N\} = ({}^t[E].[E]).\{A\}$$
$$({}^t[E].[E])^{-1}.({}^t[E].\{N\}) = \{A\} \qquad (3)$$

Whatever the number n of points, the multiplication $^t[E].[E]$ has always the dimensions (3*3), and can be easily inverted (assuming det($^t[E].[E]$) is non zero).

$$\left({}^t[E]\cdot[E]\right)=\begin{bmatrix}\sum_{i=1}^{n}\varepsilon_1^{i2} & \sum_{i=1}^{n}\varepsilon_1^i\varepsilon_2^i & 0\\ \sum_{i=1}^{n}\varepsilon_1^i\varepsilon_2^i & \sum_{i=1}^{n}\left(\varepsilon_1^{i2}+\varepsilon_2^{i2}\right) & \sum_{i=1}^{n}\varepsilon_1^i\varepsilon_2^i\\ 0 & \sum_{i=1}^{n}\varepsilon_1^i\varepsilon_2^i & \sum_{i=1}^{n}\varepsilon_2^{i2}\end{bmatrix} \quad \text{more simply} \begin{bmatrix}a & d & 0\\ d & b & d\\ 0 & d & c\end{bmatrix} \qquad (4)$$

$$\left({}^t[E]\cdot[E]\right)^{-1}=\frac{1}{abc-d^2(a+c)}\begin{bmatrix}bc-d^2 & -cd & d^2\\ -cd & ac & -ad\\ d^2 & -ad & ab-d^2\end{bmatrix} \qquad (5)$$

Then, the matrix [A] can be determined with equation (3).

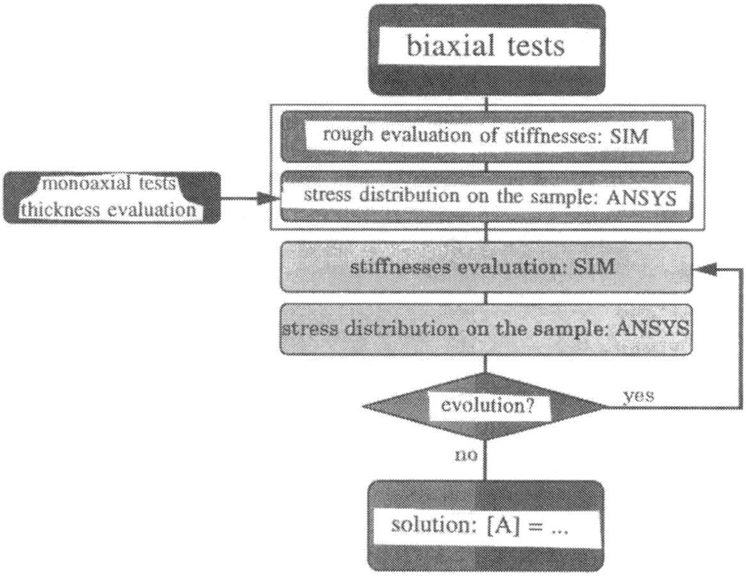

The procedure is as follow:
First, we carry out a set of three biaxial tests with the different loading ratios. Using SIM, we are able to make a first rough evaluation of the stiffnesses, assuming that the stresses are homogeneous . With these results and information given from monoaxial tests and the measurement of the material thickness we can make a first evaluation of the stresses distribution on the sample. Now considering the new stresses in the central part of the sample, and using the SIM, we can calculate the A matrix... And so on until the stabilization of the results.

4.2. STRESS DISTRIBUTION

First of all, we must specify that the stresses given by load cells at the jaw level do not correspond to these in the central part of the sample. Actually, the sample at the jaw level is only submitted to a monoaxial tension, whereas in its very central part it is submitted to a biaxial tension. Thanks to the software ANSYS, we tried to bring out the stress distribution on the sample. The sample has been meshed with Shell63, 4-noded elastic shell element for which large deflection is included and having a 'membrane only' capability as an option.

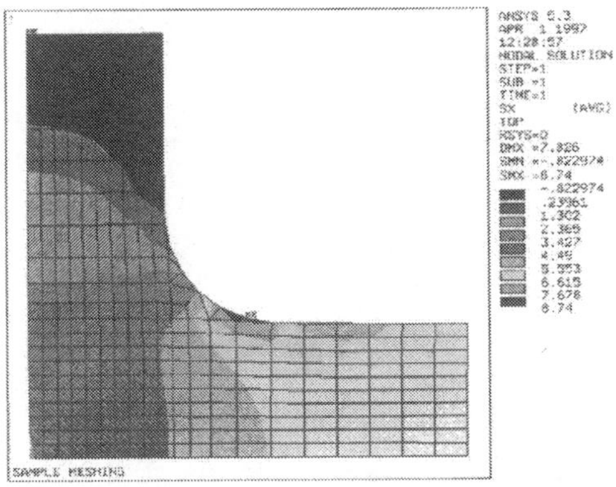

Figure 5. Stress distribution on the sample.

4.3. APPLICATION

We mainly studied a quasi-linear zone of the global behaviour (Figure 6). It corresponds to the stress range applied to the fabric in place. We cannot directly use the SIM, before, we have to carry out a translation of the strain axe, that leads to the determination of N° values.

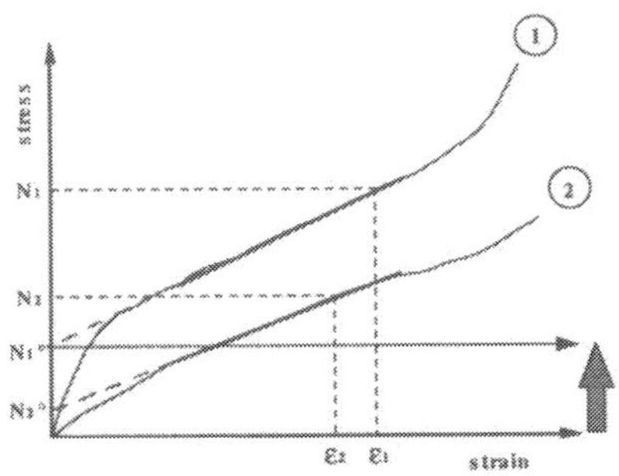

Figure 6. Stress-strain curves for both directions.

Then, the in-plane stiffness components Aij are calculated with the method presented in §4.1. Three tests were made on three different samples with a basic speed of V= 10 N/s, and using the following loading ratios: 1/1, 1/2 and 2/1.

$$\left.\begin{array}{l} N_1 = A_{11} \cdot \varepsilon_1 + A_{12} \cdot \varepsilon_2 + N_1^0 \\ N_2 = A_{12} \cdot \varepsilon_1 + A_{22} \cdot \varepsilon_2 + N_2^0 \end{array}\right\} \Rightarrow \left\{\begin{array}{l} N_1 - N_1^0 = A_{11} \cdot \varepsilon_1 + A_{12} \cdot \varepsilon_2 \\ N_2 - N_2^0 = A_{12} \cdot \varepsilon_1 + A_{22} \cdot \varepsilon_2 \end{array}\right. \qquad (6)$$

$$\begin{Bmatrix} N_1 - N_1^0 \\ N_2 - N_2^0 \end{Bmatrix} = \begin{bmatrix} A_{11} & A_{12} \\ A_{12} & A_{22} \end{bmatrix} \cdot \begin{Bmatrix} \varepsilon_1 \\ \varepsilon_2 \end{Bmatrix} \tag{7}$$

Table 1. Results obtained by association of different biaxial tests, V= 10 N/s.

TEST	1/2		2/1	
1/1	$[A]=\begin{bmatrix}82.5 & 1.8 \\ 1.8 & 118.2\end{bmatrix}$	Ewa = 138 MPa Ewe = 197 MPa $\nu_{12} = 0.015$	$[A]=\begin{bmatrix}153 & -29.9 \\ -29.9 & 136.9\end{bmatrix}$	Ewa = 244 MPa Ewe = 218 MPa $\nu_{12} = -0.23$
1/2			$[A]=\begin{bmatrix}134.4 & -11.1 \\ -11.1 & 112.1\end{bmatrix}$	Ewa = 222 MPa Ewe = 185 MPa $\nu_{12} = -0.1$
1/1 with 2/1	$[A]=\begin{bmatrix}126.2 & -12.4 \\ -12.4 & 122\end{bmatrix}$	Ewa = 208 MPa Ewe = 201 MPa $\nu_{12} = -0.10$		

[A] are given in N/mm.

5. Numerical Modelling

Our global analysis was proposed as opposed to classical methods based on a multi-scale study of the material. These methods consist in determining the characteristics of each component with mechanical tests. Then the global characteristics are calculated using different modelling.

Geometrical data were obtained by means of a micrography. The yarn material characteristics were obtained by dissolving the matrix and processing tensile tests on the yarns. The total coated fabric thickness is 0.6 mm and the global fibres ratio is 24%.

Table 2. Micrography components in warp and weft directions.

	a (mm)	b (mm)	c (mm)	DH (mm)
Cutting in warp dir.	0.912	0.057	0.194	1.223
Cutting in weft dir.	0.950	0.175	0.175	1.117

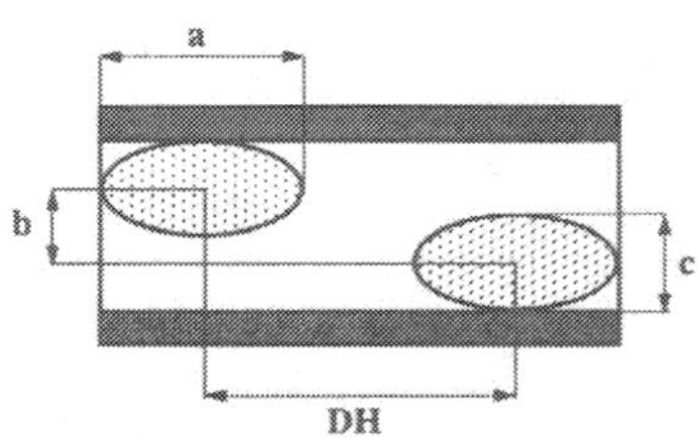

Figure 7. Woven fabric micrography diagram.

Table 3. Yarn and matrix characteristics.

	Yarns warp dir.	Yarns weft dir.	PVC matrix
Young's modulus (MPa)	3290	2530	4
Shear modulus (MPa)	1260	970	1
Poisson's ratio	0.3	0.3	0.5

5.1. THE MULTILAYER MODEL

The woven coated fabric is represented by a stacking of unidirectionally reinforced layers directed either in 0 or in 90 degrees, M.Manera [5], (Figure 8). Elastic characteristics of each layer are calculated by the law of mixtures (micromechanic). Then the global stiffnesses are determined by applying the laminated thin plate theory (macromechanic), assuming: a perfectly elastic behaviour, small strains, and a perfect adhesion between yarns and matrix as well as between layers.
The software Multicouche L2M including this facility has been developed by P. Bonnel [6] in 1990. It corresponds to a more complete approach to the software "Think Composites" developed by S.W. Tsaï and al.[7].

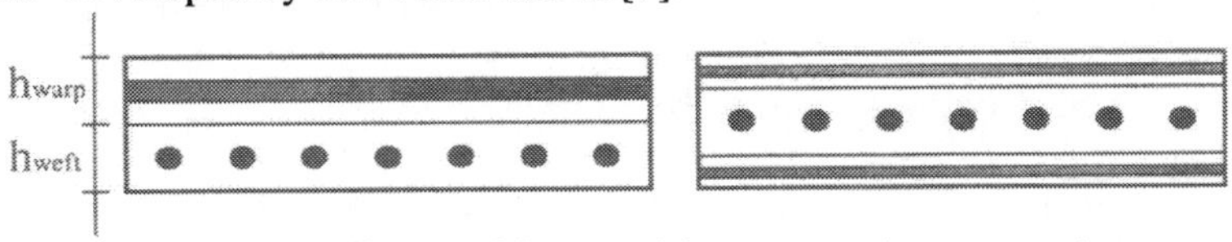

Figure 8. Two different multilayer models (a-symmetric or symmetric).

Table 4. Results obtained with multilayer models

$h_{warp} = h_{weft} = 0.3$ mm	$h_{warp} = 0.315$ mm and $h_{weft} = 0.285$ mm
$[A]=\begin{bmatrix} 240.2 & 1.7 & 0 \\ 1.7 & 185.5 & 0 \\ 0 & 0 & 0.8 \end{bmatrix}$	$[A]=\begin{bmatrix} 252 & 1.7 & 0 \\ 1.7 & 176.4 & 0 \\ 0 & 0 & 0.8 \end{bmatrix}$
Ewa = 400 MPa	Ewa = 420 MPa
Ewe = 309 MPa	Ewe = 294 MPa
$\nu_{12} = 0.01$	$\nu_{12} = 0.01$

[A] are given in N/mm.

5.2. THE CRIMP MODEL

This modelling was first developed by Ishikawa and Chou [8] in 1982. Weft yarns are represented by a sinusoidal surface which crimps between warp yarns regarded as rectilinear (Figure 9). Linear elastic stiffnesses are calculated by application of the laminated thin plate theory, first to a thin slice dx, and subsequently to the complete cell by assembly.

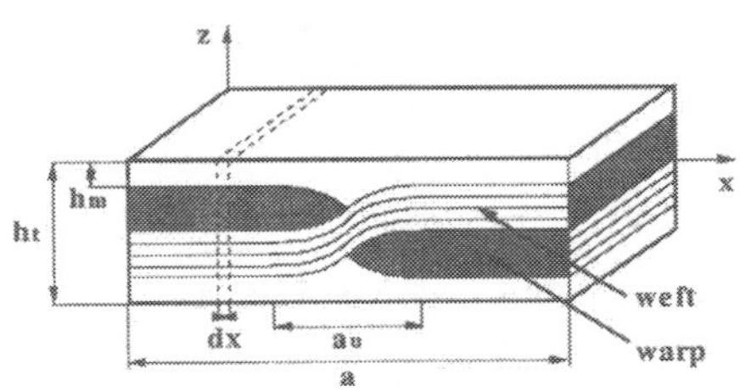

with:

$h_t = 0.6$ mm
$h_m = 0$

Figure 9. The Crimp model.

Calculations are made with the software Tissu L2M, developed by D. Genin and al [9] and which includes this model.

Table 5. Results obtained with the crimp model.

$a/a_u = 5$	$a/a_u = 1000$
$[A]=\begin{bmatrix}168.2 & 65.2 & 0\\ 65.2 & 229.1 & 0\\ 0 & 0 & 34.9\end{bmatrix}$	$[A]=\begin{bmatrix}241.4 & 40.9 & 0\\ 40.9 & 287.5 & 0\\ 0 & 0 & 32.8\end{bmatrix}$
Ewa = 340 MPa	Ewa = 468 MPa
Ewe = 249 MPa	Ewe = 393 MPa
$\nu_{12} = 0.29$	$\nu_{12} = 0.14$

[A] are given in N/mm.

The first two models consisted of a stiffnesses estimation based on the laminated plate theory. Conversely the third model is based on a geometrical approach to the problem.

5.3. PEIRCE'S AND SINE MODELS

The yarns are represented by a surface defined by a parametrical path combined with an elliptical or lenticular base. Then the cell is built by a carefully chosen positioning of the yarns, trying to avoid overlapping.
The two models differ from the path nature:

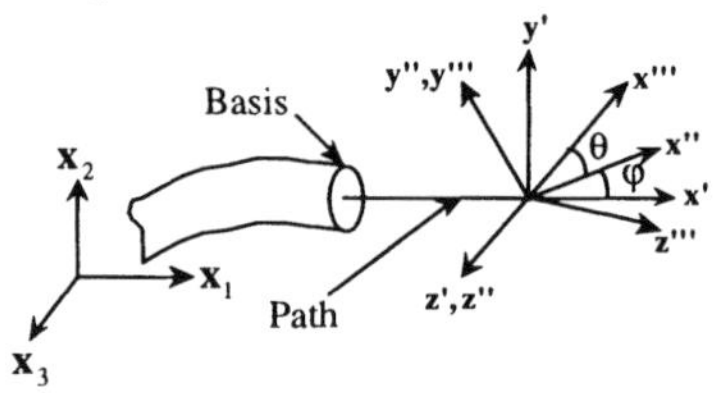

Figure 10. Peirce's and sine models.

<u>Peirce's</u>
It was established by F.T. Peirce [10] in 1937. The path is divided: a rectilinear part between two elliptical ones.

<u>sine</u>
It is proposed by P. Bonnel and D. Bigaud [11]. The path is totally sinusoidal.

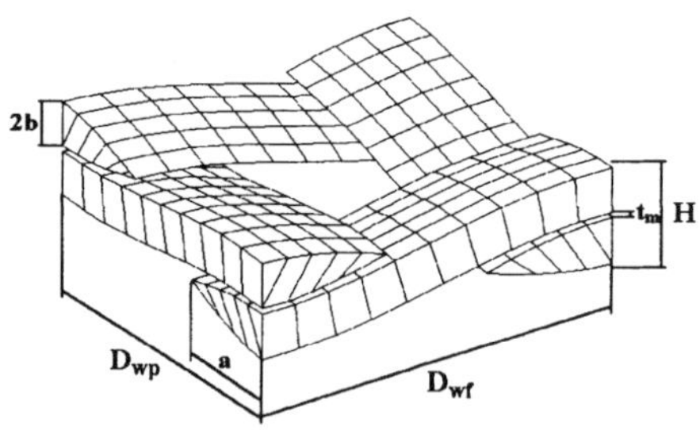

With:
a = 0.4655 mm
b = 0.09225 mm
Dwp = 1.117 mm
Dwf = 1.223 mm
H = 0.369 mm
mesh: 10(X dir.)*5(Y dir.)*10(Z dir.)

Figure 11. Dimensions for Peirce's and sine models.

The cell is subdivided by a 3D meshing. The characteristics of each element are determined according to the rate and orientation of fibres. The global stiffness of the complete cell is calculated thanks to an energetical approach.

These two models are close to the real geometry of fabrics, and allow numerous and various representations in 2D as well as in 3D. They are integrated in a software TIS3D developed by D. Bigaud.

Table 6. Results obtained with Peirce's and sine models.

Peirce's	sine
Ewa = 286 MPa	Ewa = 279 MPa
Ewe = 217 MPa	Ewe = 298 MPa
$\nu_{12} = 0$	$\nu_{12} = 0.074$

6. Conclusion

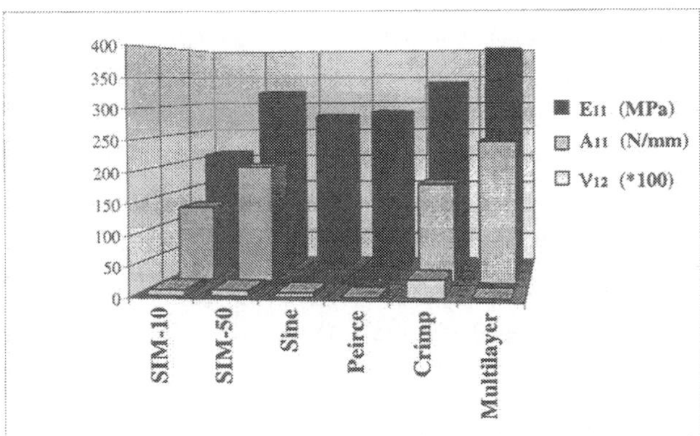

Figure 12. Stiffness characteristics comparison.

Mixed numerical and experimental methods for mechanical identification became self - sufficient thanks to the development of computering and experimental means. The classical methods require specific and complex techniques such as micrography with micro-measurements, matrix dissolving and tensile tests on the yarns, while our method requires tests on the complete structure at a macro-scale. Moreover, the results given as a stiffness matrix could be directly used by designers in calculation codes. A good improvement for our method would be to get rid of the thickness measurement and the Young's moduli in order to work exclusively on the in-plane stiffness components [A]. This should be done thanks to a more extensive use of the numerical means.

7. References

1- Mailler, P., Hamelin, P., Jarrigeon, M. and Nemoz, G., *Comportement en grandes déformations des matériaux composites souples soumis à un chargement biaxial*, in 'Annales des composites', 1995/3, Essais multiaxiaux et composites, Journées AMAC, Besançon, january 1995, Ed. AMAC.

2- Zhao, Y.H., *Visualisation du champ de déformation de surfaces textiles. Mise au point et utilisation d'un logiciel d'analyse d'image*, Ph. D. thesis, Ecole Centrale Paris, 1991.

3- Mailler, P. and Hamelin, P., *Methodology of geometrical modeling and theoretical and experimental stiffness characterization: applied to textile reinforced composites structures*, 2nd International Symposium Texcomp-2, 'New Textiles for Composites', Leuven, Belgium, May 1994, Ed. Verpoest and Ko.

4- De Wilde, W.P., *Numerical modelling of linear elastic and viscoelastic response of composite structures*, Vrije Universiteit Brussel, Belgium, 1995.

5- Manera, M., Massot, J.J. and Morel, G., *Manuel de calcul des composites verre-résine*, (Ed. Pluralis, Paris 1988).

6- Bonnel, P., *Modélisation géométrique et méthodes théorico-expérimentales d'identification des rigidités de composites à renforts textiles*, Ph. D. thesis, Université Claude Bernard Lyon I, 1994.

7- Tsaï, S.W., *Composites design*, 3rd edn (Think Composites, Dayton, USA, 1987).

8- Ishikawa, T. and Chou, T.W., *Elastic behavior of woven hybrid composites*, Journal of Composite Materials, vol 16, n°1, 1982.

9- Genin, D., Hamelin, P. and Chevalier, Y., *Interactions chaîne-trame dans les composites renforcés par tissus*, in 'Comportement des composites à renfort tissus', colloque du GRECO 92, may 1988, Franc, Ed. Pluralis 1990.

10- Peirce, F.T., *The geometry of cloth structure*, Journal of the Textile Institute, vol 28, 1937, pp 81-88.

11- Bigaud, D., Bonnel, P. and Hamelin, P., *TIS3D 2.0: A geometric modeling software for reinforced textile composites*, International Symposium Euromech 334, 'Textile Composites and Textile Structures', Lyon, France, May 1995.

A SIMPLE METHOD FOR DETERMINING THE ANISOTROPIC PROPERTIES OF COLD DRAWN WIRE

B. CARLSSON
Dept. of Metal Forming
Royal Institute of Technology, Stockholm

1. Introduction

Metal wire produced by cold drawing often shows a high degree of anisotropy. If the wire is formed plastically after the drawing operation, the anisotropy will influence the deformation and the final geometry of the end product. However, the anisotropy of cold drawn wire can often be approximated to planar isotropy, that is, the properties are isotropic in the plane the normal of which is parallel to the drawing direction.

If planar isotropy can be assumed, which can be verified through hardness tests across the diameter of the wire, the constitutive equation simplifies significantly. If Hill's 1948 criterion is used, only three independent material parameters have to be determined, instead of six in the general case.

A method has been developed to determine the material properties for a wire which exhibits planar isotropy. The method is based on inverse simulation of certain deformations of the wire. The flow stress of the wire together with the constants in the Hill 1948 criterion are determined. The inverse simulations have been performed with the commercial finite element code Abaqus. The method is simple and is thought to be suitable for use in industry.

2. Background

Metal wire of small dimensions is usually produced by cold drawing. In the process, a texture will develop in the material and the resulting wire will be anisotropic to some degree. Due to the symmetry of the drawing operation the anisotropy is likely not to show any variation with the circumferential angle. However, a radial variation is sometimes seen.

Cold drawn wire is a semi- manufacture which is used in a variety of end products. In many cases the wire is formed plastically after cold drawing and the deformation behaviour in such a forming is influenced by the anisotropy. A knowledge about the anisotropy is required to predict the material response.

H. Sol and C.W.J. Oomens (eds.), Material Identification Using Mixed Numerical Experimental Methods, 51-60.

The properties of anisotropic sheet metal are determined through tension tests. The deformation in different directions are measured and from this the anisotropic properties are calculated. This is not applicable to metal wire due to the small dimensions. It is not possible to cut out tensile test specimens in different directions from a wire without altering the structure of the material. Even if it was possible it would be an awkward way of determining the material properties.

One effect of the anisotropy is the preferred deformation in certain directions. This is used in sheet metal testing, where the directions of the principal strains are known in advance. It can be said that the determination of the anisotropic properties of sheet metal is a very simple case of inverse simulation, since the properties are not directly measured but are obtained from the strain ratios.

This idea of inverse simulation can be generalized. Theoretically, the material parameters in an anisotropic yield criterion can be determined by as many tests as there are parameters in the criterion. The tests must contain significant deformations in the directions affected by the anisotropic parameters. The tests can be simulated with the use of a finite element code, under variation of the material parameters, until the best fit to test results is obtained. In this case, both measured deformations and forces from the tests can be used to simulate against.

The method presented here is an improvement of an earlier method [1].

3. Description of Method

3.1 YIELD CRITERION

A yield criterion and associated flow rule has to be chosen for the description of a plastically deforming metal. One of the earliest and most used is Hill's 1948 anisotropic criterion [2]. It is an extension of the Huber- Mises criterion and consists of six material parameters (in addition to Huber- Mises) describing the degree of anisotropy with respect to the coordinate axes. The criterion is written

$$\begin{aligned} 2f(\sigma_{ij}) = F(\sigma_y - \sigma_z)^2 + G(\sigma_z - \sigma_x)^2 + H(\sigma_x - \sigma_y)^2 \\ + 2L\tau_{yz}^2 + 2M\tau_{zx}^2 + 2N\tau_{xy}^2 = 1 \end{aligned} \tag{1}$$

Hill's 1948 criterion is included as a standard in most finite element codes. For the case of planar isotropy, Hill's 1948 criterion simplifies considerably. If planar isotropy is assumed in the xy- plane it can be shown [3] that F = G = N - 2H and L = M.

Hill's 1948 criterion does not assume any variation of the anisotropy with deformation. Therefore, the flow rule associated with the criterion predicts constant anisotropy ratios. Thus, the anisotropy can be described by the ratios to a single flow stress curve.

Assume, for instance, that the flow stress curve in the z- direction is known for a material that exhibits planar isotropy in the xy- plane. By insertion in eqn (1) the flow stress in the z- direction is 1/(2N-4H). The ratio of the flow stress curve in the x- or y- direction to the flow stress curve in the z- direction is then (2N-4H)/(N-H). According to the Huber- Mises criterion, the uniaxial yield stress is related to the shear yield stress as $\tau = \sigma / \sqrt{3}$ and therefore the ratio of the xy- shear stress to the z- direction flow stress is $\sqrt{3}(2N-4H)/N$. The ratios of the shear stresses in yz- and zx- directions are $\sqrt{3}(2N-4H)/L$.

3.2 DEGREE OF PLANAR ISOTROPY

The hardness of a material, determined through a Vickers test, is proportional to the flow stress [4]. A series of Vickers indentations across the diameter of an annealed wire therefore gives an indication of the degree of planar isotropy.

The material in the present investigation is an annealed cold drawn wire of austenitic stainless steel, AISI 201, with an initial diameter of 1.61 mm. Vickers indentations were performed on the wire and the results are shown in figure 1. It is seen that the radial variation in hardness is small and, therefore, the conclusion is made that the wire can be assumed to show planar isotropy.

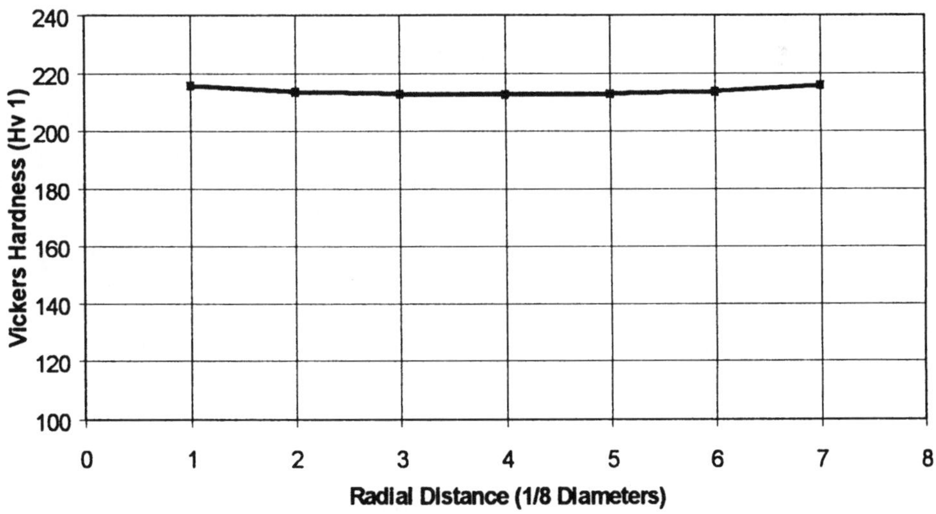

Figure 1. Vickers hardness (Hv 1) over the cross- section of the wire.

3.3 DETERMINATION OF HILL'S PARAMETERS

If the flow stress curve in one direction is known, only two of the parameters N, H and L have to be determined to completely describe a planar isotropic material, since the third parameter is obtained from the flow stress. A simple device according to figure 2 is used for this purpose. The device consists of an ordinary screw press with a load cell connected to the anvil, which is seen in figure 3.

The wire is deformed in two different ways in the press and the load is recorded. Inverse simulations are then performed until the best fit between experiments and simulations are obtained.

3.3.1. *Flat compression*

The first deformation consists of a so- called sidepressing of a wire specimen. Pieces of wire are prepared in ten millimeter lengths. A specimen is placed on the anvil which is covered with one layer of teflon tape on each face. The screw is tightened under continuous monitoring of the load, until a small plastic deformation has occured. The screw is then loosened and the specimen is taken out for measurement of the deformed height and width. Measurements are done with a 1/1000 mm accurate micrometer.

New teflon tape is applied to the anvil and the process is continued until the desired height reduction of the wire is reached. The procedure is repeated ten times for each wire type to get some statistical safety. The results are a curve of the applied load versus specimen height and a curve of the specimen width versus specimen height.

It could be argued that the specimen height should be measured when the load is still applied, since the measured load also is caused by elastic deformation. The elastic deformation of the wire specimen, however, is so small that the error can be ignored.

The friction can be assumed constant during the test since the teflon layer is renewed several times. Investigations of the magnitude of the coefficient of friction for plane strain compression tests have been done with the same teflon tape [5]. A relatively constant value of 0.04 was obtained. The coefficient of friction is therefore taken to be constant and equal to this value in all tests.

3.3.2. *45^0 compression*

The second deformation is done with a separate tool and at 45^0 to the length axis of the wire. A narrow wedge- shaped tool is used. The deformed geometry (maximum width) of the specimen is recorded as a function of compressed height.

3.3.3 *Inverse simulations*

Two finite element models simulating the wire deformations have been put together using the code Abaqus [6]. Only one eight of the wire needs to be modelled in the flat compression because of symmetry. Half the wire is modelled in the 45^0 compression. The material model is Hill's 1948.

FIGURE 2. Screw press used for the tests.

FIGURE 3. Anvil with load cell, test specimen and teflon tape.

Simulations of the compressions are done with varying anisotropic parameters until the best fit with experiments is achieved. The first round of inverse simulations are performed with a flow stress curve taken from literature. No comparison is made with the measured compressive load in this round.

When the best fit is achieved, the second round of simulations is done. A comparison between the measured and the simulated load curve is the purpose of this round. The yield stress curve in the simulations is altered until reasonable agreement is obtained between the experimental and the simulated load-displacement curve. Hill's parameters only need to be slightly modified in this round of simulations. The process is continued until load- displacement curves and deformed geometry- displacement curves agree with experimental results.

3.3.4. *Co- rotational formulation*

When performing the inverse simulations the material definitions must be made in a co- rotational frame. This is easily done in the finite element code Abaqus by using the *ORIENTATION command, which means that the anisotropy is defined with respect to the material coordinates instead of the spatial coordinates.

4. Results

An example of deformed test specimens is given in figure 4. The reduction in height in both cases was about 37%.

The wire material exhibited a very small radial variation in hardness, see figure 1, and the assumption of planar isotropy therefore seemed justified, as discussed in section 3.2.

The measured load and the simulated load, with the flow stress inversely determined from the test, are shown in figure 5. The lengthwise flow stress was fitted to a Ludwik expression, $\sigma = K_0 + K_1 \varepsilon^n$.

The measured width of the wire during the flat compression is shown in figure 6, together with the simulated width. The maximum width in the 45^0 compression is shown in figure 7 together with simulations.

A summary of the inversely determined material parameters is given in table 1. Results from a conventional tension test of the wire are also shown.

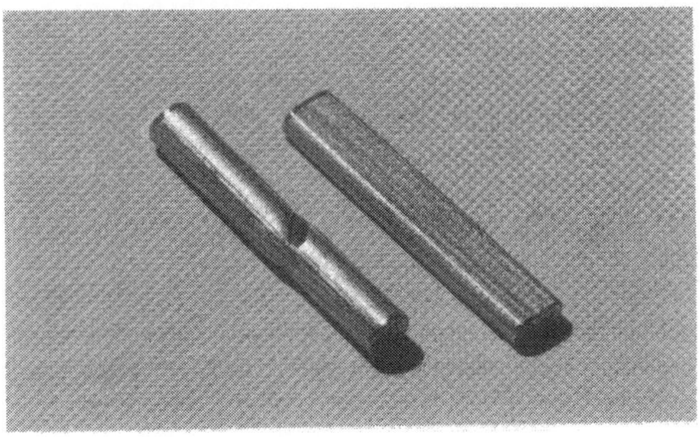

FIGURE 4. Flat and 45^0 test specimens.

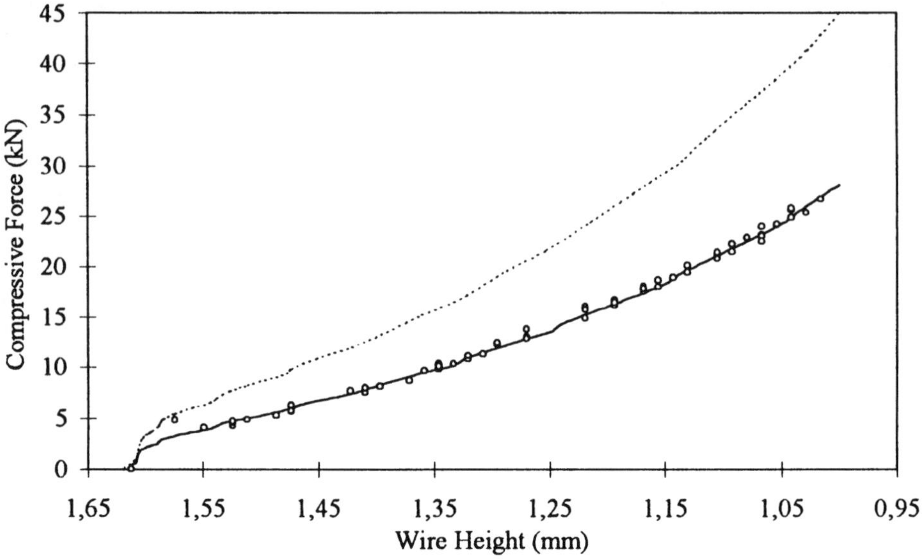

FIGURE 5. Measured (o) and simulated (solid line) compressive force as a function of instantaneous wire height for flat compression. Isotropic (dotted line) simulation shown for comparison.

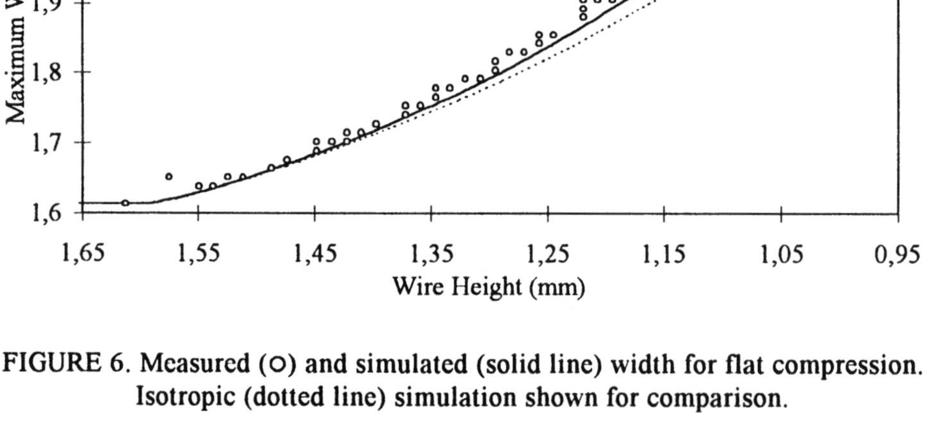

FIGURE 6. Measured (o) and simulated (solid line) width for flat compression. Isotropic (dotted line) simulation shown for comparison.

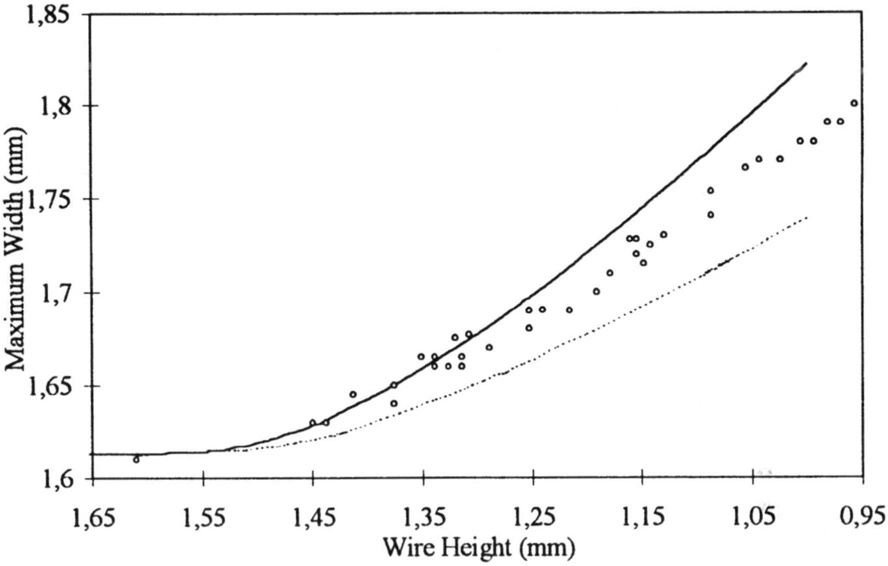

FIGURE 7. Measured (o) and simulated (solid) maximum width for 45^0 compression. Isotropic (dotted line) simulation shown for comparison.

TABLE 1. Summary of material parameters determined with the current method.

Constants in Ludwik's expression determined through inverse simulation (lengthwise flow stress):

K_0	K_1	n
280	2300	0.8

Constants in Ludwik's expression determined through tension test (lengthwise flow stress):

K_0	K_1	n
310	1850	0.83

Summary of Hill's 1948 parameters (normalized against inversely determined flow stress):

F	G	H	L	M	N
0.5	0.5	1.5	1.9	1.9	3.5

5. Conclusions

The presented method to determine the anisotropic properties of wire is limited to planar isotropy. It is, however, not unusual that cold drawn wire of small diameter exhibits planar isotropy.

The poor agreement with experiments is probably caused by the inherent inaccuracies of Hill's 1948 criterion. The use of a better anisotropic criterion is required if reliable predictions are to be made. A possible explanation for the lack of agreement with theory could be the occurence of a martensitic transformation, which would change the ratios of the flow stress curves. Nevertheless, the method is general and can be used with anisotropic criteria other than Hill's 1948.

Together with a proper anisotropic criterion, the method is thought to be suitable for use in industry due to it's simplicity and low cost.

References

1. Carlsson, B., Huml, P.: Determination of the Material Properties of an Anisotropic Metal Wire, *Annals of the CIRP* **45/1** (1996), 231-33.

2. Hill, R. : A Theory of Yielding and Plastic Flow of Anisotropic Metals, *Proc. Roy. Soc.* **193A** (1948), 281-97.

3. Hosford, W.: *The Mechanics of Crystals and Textured Polycrystals,* Oxford University Press, New York, 1993.

4. Kunda, B.: The Distribution of Hardness in the Cross- Section of Steel Wires, *The Wire Industry* Dec. 1965, 1178-82.

5. Zhang, J., Huml, P.: Evaluation of Friction in Plane Compression Test, *J. Mater. Proc.* (in press)

6. Hibbit, Karlsson & Sorensen, Inc.: Abaqus Manuals, Ver. 5.5

ANALYSIS OF INHOMOGENEOUS DISPLACEMENT FIELDS FOR THE IDENTIFICATION OF PARAMETERS FOR ELASTO-PLASTIC DEFORMATION LAWS

U.-J. GÖRKE, S. KRETZSCHMAR AND R. KREIßIG
Technische Universität Chemnitz-Zwickau,
Institut für Mechanik
Straße der Nationen 62, D-09111 Chemnitz

1. Introduction

The design and development stages of the production process have increasingly changed using numerical methods for strength analysis and the optimization of components and structures. Costly experimental test series have been superseded by more economical numerical calculations the results of which can be experimentally verified on optimized prototypes. Especially the Finite Element Method (FEM) has been established in the industrial practice as an effective numerical aid for engineers. The results of the numerical simulation of the mechanical behaviour essentially depend on the implemented material laws which describe the real material behaviour sufficiently exactly.

Balance laws valid for all materials in the sense of the theory of the classical mechanics of continua were formulated taking into consideration physical experience. These relations form the well-known system of fundamental laws of the mechanics of continua.

In components consisting of different materials different stress-strain states could be determined under identical load conditions which is not taken into account by the balance laws. For describing these observations we have used equations establishing correlations between variables typical of the special constitution of the material. These equations are known as material laws (constitutive equations) or deformation laws if they contain only mechanical variables.

The modelling of the material behaviour starts with the definition of a suitable physical model restricted to fundamental material properties.

H. Sol and C.W.J. Oomens (eds.), Material Identification Using Mixed Numerical Experimental Methods, 61-70.

From experience constitutive principles and axioms arise, facilitating the consistent formulation of material laws [1].

Based on these principles special material laws can be formulated as the mathematical idealization of the real material behaviour (structural optimization). In this paper the behaviour of elasto-plastic materials with isotropic and kinematic hardening will be described, regarding small deformations characterized by the following relations:

- Relation between the tensors of stress velocity $\dot{\boldsymbol{\sigma}}$ and strain velocity $\dot{\boldsymbol{\varepsilon}}$

$$\dot{\boldsymbol{\sigma}} = \boldsymbol{f}_{\sigma}(\boldsymbol{\sigma}, \boldsymbol{\kappa}, \boldsymbol{p})\, \dot{\boldsymbol{\varepsilon}} \tag{1}$$

- Evolutional equations for internal variables $\boldsymbol{\kappa}$ (tensorial quantities describing the hardening behaviour)

$$\dot{\boldsymbol{\kappa}} = \boldsymbol{f}_{\kappa}(\boldsymbol{\sigma}, \boldsymbol{\kappa}, \boldsymbol{p})\, \dot{\boldsymbol{\varepsilon}} \tag{2}$$

- Yield condition

$$F(\boldsymbol{\sigma}, \boldsymbol{\kappa}, \boldsymbol{p}) = 0. \tag{3}$$

The variables in equations (1)-(3) are connected with each other via material parameters $\boldsymbol{p} = (p_1, \ldots, p_{n_p})^T$. The identification of these quantities (parameter optimization) using experimental results completes the process of material modelling and has an important effect on the quality of numerical results simulating the mechanical behaviour of components and structures.

2. Methods for the Identification of Material Parameters

As a rule material parameters cannot be determined directly because they do not represent physical quantities ascertainable by measurement methods. They may be identified by their effect on stress-strain states which for their part cannot be measured in the case of an inhomogeneous distribution but result from variations of displacements, forces and momentums. Therefore, material parameters represent the "*reason*" for the evolution of deformation fields during loading processes. The calculation of deformation fields with given material parameters is known as the direct problem. The identification of material parameters analyzing their "*effects*" in the form of displacement fields is an inverse problem.

Direct and inverse problems can be well-posed or ill-posed. In accordance with a definition of Hadamard [2] a problem is well-posed if its solution exists and if it is unique and stable. As a rule the problem of the identification of parameters for the nonlinear material behaviour is an ill-posed problem. The mapping operator connecting material parameters with displacement fields is a nonlinear implicit one, and its structure is not known. Therefore its inversion is impossible, and the problem of parameter identification was formulated as the problem of adjusting the parameters to minimize a cost function

$$f(\boldsymbol{p}) = \sum (\boldsymbol{u}(\boldsymbol{p}) - \bar{\boldsymbol{u}})^2 \longrightarrow \min_{\boldsymbol{p} \in P} \tag{4}$$

until the calculated displacements $\boldsymbol{u}(\boldsymbol{p})$ match the measured ones $\bar{\boldsymbol{u}}$ in the sense of a least square formulation.

The optimization problem (4) is not necessarily a well-posed one, especially if the cost function is not convex. The reasons for the possible ill-posedness of the parameter optimization problem are the physically and mathematically idealized nature of the description of the real material behaviour by deformation laws and the incompleteness of readings which in addition may be incorrect (disturbances, scattering). To reduce the ill-posedness of the problem of parameter identification it is necessary to obtain a maximum of information about the material behaviour and to take it into account in the material model. Nevertheless the nonuniqueness and instability of the solution of optimization problem (4) is not excluded, but every existing solution is an acceptable solution in accordance with the initial information.

A commonly used method for material parameter identification is the analysis of homogeneous stress-strain states which are produced in measurement regions of specially formed specimens applying special loading paths (tension, compression, torsion etc.). Stresses and strains can be calculated directly from applied forces and displacements using geometrical proportions of the specimen. Minimization of a cost function analyzing homogeneous states in many cases results in the numerically simple problem of solving a system of linear equations. Otherwise this method has essential disadvantages which reduce its practicability for parameter identification. The generation of homogeneous stress-strain states is a demanding experimental problem, and the deformation laws are frequently not suitable to describe complex inhomogeneous stress-strain states, using the determined parameters.

The experimental generation of inhomogeneous displacement fields is comparatively simple, and their numerical analysis permits the consideration of points with different load histories in the cost function (4) which increases the information set for parameter identification. Because of the necessity of the solution of the direct (field) problem to determine the calculated displacement values at each iteration step of the optimization process, this method is very time-intensive.

In the following sections we will represent the results of the application of nonlinear optimization methods in determining hardening parameters of elasto-plastic deformation laws, analyzing inhomogeneous displacement fields of a notched four-point bending specimen [3, 4]. The direct problem was solved using the FEM-program PMEP developed by the research group "Scientific Parallel Computing" at the Chemnitz University of Technology [5]. For the solution of the inverse problem and sensitivity analysis we have chosen efficient optimization algorithms in accordance with results published in the literature [6, 7]. All calculations were made on a ParsytecGC/PowerPlus-128 Supercomputer with 128 processors. The solution time for an average optimization problem (9 load increments in the direct problem, 5...8 optimization steps) was between 30 and 40 minutes.

3. Analysis of the Displacement Field of a Notched Four-point Bending Specimen

The four-point bending test has some advantages for parameter identification:

- Simple realization of the plane-stress state,
- Determination of different load histories for several material points on the surface of the specimen (parameter identification on average),
- Simple realization of loading cycles turning about the specimen to obtain the necessary information for the identification of different kinds of hardening.

Figure 1 shows the geometry of the specimen used.

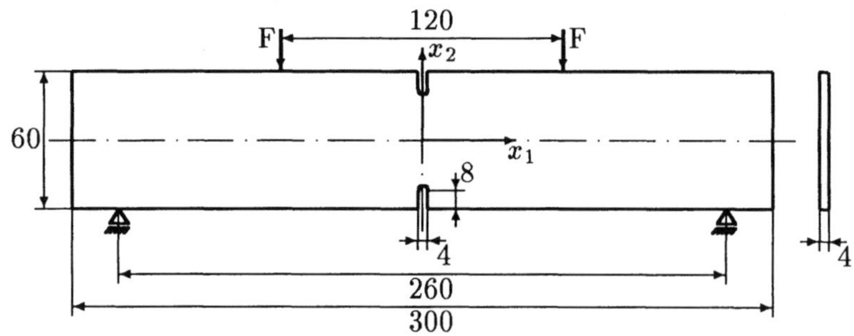

Figure 1. Four-point bending. Geometry of a notched specimen and loading conditions.

3.1. MEASUREMENT AND APPROXIMATION OF IN-PLANE DISPLACEMENT VALUES

For the measurement of the in-plane displacements u_x and u_y at defined levels of the loading cycle the Moiré method was used. Upper light patterns, the so-called isothetes (Fig. 2), were obtained by putting the deformed raster screen on an undeformed initial screen. They represent contour lines of displacement fields u_x and u_y for the measured loading step.

Measured and calculated displacement values are compared at the nodes of the FE-mesh to identify material parameters, using an optimization method. In general, measurements detected on the isothetes are not available at the needed points. Therefore, the experimental results are approximated, using special isoparametric eight-node quadrilateral elements to minimize a function of the least-square type [4]. Figure 2 represents the good agreement of measured and approximated contour lines of displacement fields. After approximation the displacement fields are transformed into the undeformed initial configuration to perform the conditions of the material description used by FEM.

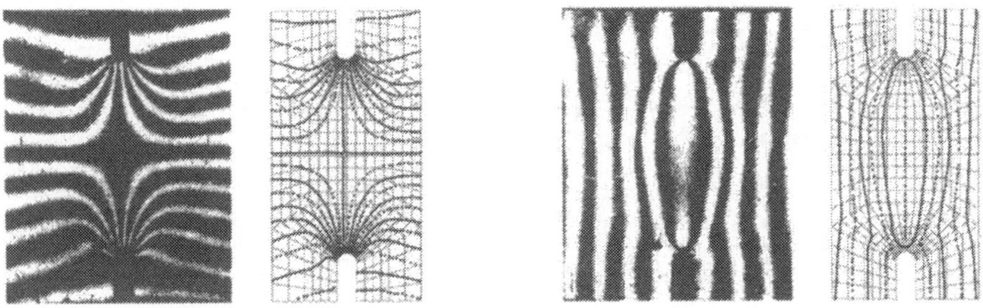

Figure 2. Four-point bending of a notched specimen. Comparison of Moiré-images and approximated values for displacements u_x (left) and u_y (right).

For the verification of the proposed identification method the further presented results are referred to *"artificially"* measured displacement fields generated by a FEM solution according to a given loading path and given material parameters.

3.2. FINITE ELEMENT MODEL

For comparing the artificially measured and calculated displacement fields an FE-model was generated for the four-point bending specimen (Fig. 3). For considering the conditions of symmetry an FE-mesh was generated for the right side of the specimen. To increase the numerical efficiency the model of the specimen was additionally shortened. For the simple realization of loading and reloading steps and the improvement of the numerical stability the load was applied as a triangularly distributed load at the right boundary.

Figure 3 also represents a typical FE-mesh resulting from the refinement of a coarse grid in two levels. This model contains 6162 degrees of freedom and was handled with the highest efficiency on eight processors.

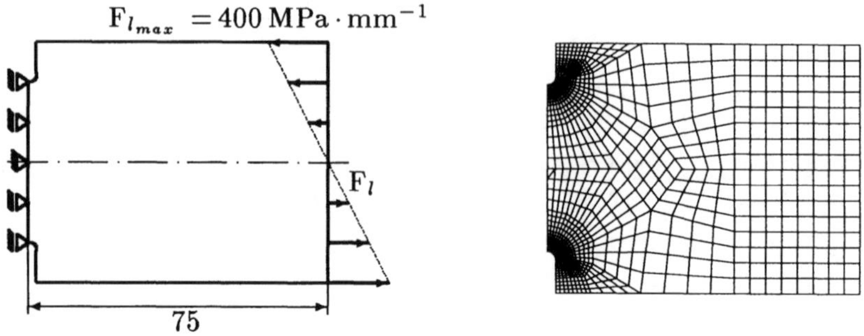

Figure 3. Finite element model of a four-point bending specimen. Geometry and boundary conditions (left). Finite element mesh of the four-point bending specimen with two-level mesh refinement of the coarse grid (976 elements, 3081 nodes).

3.3. FORMULATION AND INTEGRATION OF THE DEFORMATION LAW FOR ELASTO-PLASTIC BEHAVIOUR

The elastic part of the employed deformation law was approximated by linear Hooke's law.

According to the description of the theory of scleronomic plastic flow given in [8] the yield function was assumed in terms of the deviatoric stresses $\overset{D}{\sigma}_{ij}$ and the back stresses α_{ij} as follows:

$$F = \frac{3}{2}\left(\overset{D}{\sigma}_{ij} - \alpha_{ij}\right)\left(\overset{D}{\sigma}_{ij} - \alpha_{ij}\right) - \sigma_F^2 = 3 \overset{\overset{D}{\boldsymbol{\sigma}}-\boldsymbol{\alpha}}{II} - \sigma_F^2 = 0\,. \tag{5}$$

For the determination of yield stress σ_F as a function of the equivalent plastic strain ε_v^{pl} we used the Ludwik-Ulbricht relation

$$\sigma_F = \sigma_{Fo} + \tilde{a}\left[\left(\varepsilon_v^{pl} + \beta\right)^n - \beta^n\right] \qquad (\tilde{a} = \sigma_{Fo} \cdot a) \tag{6}$$

and for the evolution of the back stresses Prager's kinematic hardening rule

$$\dot{\alpha}_{ij} = c\,\dot{\varepsilon}_{ij}^{pl} = c\,\dot{\lambda}\,\frac{\partial F}{\partial \sigma_{ij}} = 3c\left(\overset{D}{\sigma}_{ij} - \alpha_{ij}\right)\dot{\lambda} \tag{7}$$

with the plastic multiplier $\dot{\lambda}$. The material was assumed to be isotropic ($\alpha_{ij}^o = 0$) in the undeformed state. The consideration of an initial orthotropy in the FE code PMEP is uncomplicated.

As the elastic parameters can be determined analyzing homogeneous stress-strain states, the solution vector for the optimization process analyzing inhomogeneous displacement fields contains the parameters $\boldsymbol{p} = (\sigma_{Fo}, \tilde{a}, n, \beta, c)^T$ characterizing the plastic yield behaviour of the material.

The deformation law describing the elasto-plastic material behaviour consists of a system of differential equations (1), (2) and an algebraic condition (3). This algebro-differential system has to be solved at each integration point of the Finite Element model after the solution of the stiffness system in each equilibrium iteration. Using a general difference method for the numerical solution of a differential equation $\dot{y} = f(t, y)$ to integrate the deformation law (1)-(3), these relations result in a system of nonlinear algebraic equations which was solved by a Newton method.

3.4. OPTIMIZATION METHOD AND SENSITIVITY ANALYSIS

The cost function to minimize with respect to the vector of material parameters $\boldsymbol{p} = (p_1, \ldots, p_{n_p})^T$ is given by

$$\Phi(\boldsymbol{p}) = \frac{1}{2} \sum_{i=1}^{n_L} \sum_{j=1}^{n_T} \left[\left(\{\hat{u}_x(\boldsymbol{p})\}_{ij} - \{\tilde{u}_x\}_{ij} \right)^2 + \left(\{\hat{u}_y(\boldsymbol{p})\}_{ij} - \{\tilde{u}_y\}_{ij} \right)^2 \right] \quad (8)$$

where $\hat{u}_x$, $\hat{u}_y$ are the calculated displacement vectors and $\tilde{u}_x$, $\tilde{u}_y$ are the measured ones [4]. Calculated and measured displacement values are compared for n_T material points of n_L loading steps. The nonlinear direct problem for the numerical approximation of the displacement field has to be solved at each optimization step.

For the solution of nonlinear optimization problems two classes of methods are known – gradientless methods and gradient based methods. Gradientless stochastic or evolutional strategies are not suitable methods for the determination of material parameters analyzing inhomogeneous displacement fields because of their large number of essential calculations of the cost function (8) (up to $10^5 \ldots 10^6$ trials to determine the optimum is not unusual).

Gradient based methods for solving optimization problems are founded on an iterative modification of a vector of material parameters $\boldsymbol{p}_{l+1}$, using the previous solution $\boldsymbol{p}_l$ [9]:

$$\boldsymbol{p}_{l+1} = \boldsymbol{p}_l + \omega\, \boldsymbol{s}_l = \boldsymbol{p}_l - \boldsymbol{Q}_l^{-1}\, \nabla\Phi_l \quad (9)$$

with search direction $\boldsymbol{s}_l$ and search step length ω. The gradient $\nabla\Phi_l$ is determined by the structure of the cost function (8), and so the search direction has been decisively affected by matrix $\boldsymbol{Q}_l$.

The simplest gradient based method is the method of the steepest descent with $\boldsymbol{Q}_l^{SD} = \boldsymbol{I}$. Using Newton's method

$$\boldsymbol{Q}_l^N = \nabla^2\Phi_l, \quad \omega = 1 \quad (10)$$

a better convergence of the solution (quadratic near the optimum) can be achieved if the matrix of the second derivatives of the cost function with respect to the material parameters, Hessian matrix, is positive definite. For the represented investigations a modification of Newton's method was used, a Gauss-Newton-type method, in which the second derivatives of the displacements with respect to the material parameters in Hessian matrix were neglected.

A combination of the method of steepest descent and Newton's method is the Levenberg-Marquardt method with

$$\boldsymbol{Q}_l^{LM} = \boldsymbol{Q}_l^N + \mu \boldsymbol{I}, \quad \omega = 1 . \tag{11}$$

Restrictions for the material parameters were not considered generating the mathematical structure of the cost function, but optimization has not been regarded as successful if only one of the parameters violates limits established due to physical considerations ($(p_i)_{min} \leq p_i \leq (p_i)_{max}$, $i = 1, \ldots, n_p$).

The objective of sensitivity analysis is the determination of the first derivatives of the displacements with respect to the material parameters included in the gradient of the cost function (8). For the results presented we used a semianalytical sensitivity analysis, using the analogical numerical structure of the processes of integration of the deformation law and the determination of the consistent stiffness matrices (for details see [4, 5]).

4. Factors Effecting the Numerical Parameter Identification

The following factors have a determining influence on the results of the identification of material parameters analyzing inhomogeneous displacement fields:

- Selection of suitable experiments,
- Modification of the Finite Element model,
- Choice of the special gradient based optimization method,
- Number and magnitude of the load steps,
- Number and initial values of the material parameters,
- Errors in measured values,
- Location of material points in which measured and calculated displacements will be compared,
- Choice of the special difference scheme for the integration of the deformation law and the sensitivity analysis.

At the Institute of Mechanics of the Chemnitz University of Technology we carried out systematic investigations to determine the influence of the named factors. Detailed results were presented in [10]. Some of the most important results will be given in the following.

a) Whereas the convergence speed of the method of steepest descent proved to be inacceptably slow, the best results have been achieved using the Levenberg-Marquardt method concerning the convergence of the optimization process and the accuracy of the expected values of the optimized parameters.

b) In many cases the expected material parameters could only be identified taking into account load steps both of the loading path and the reloading path. Under the consideration of exclusively using load steps of the loading history the problem of mutual dependency of some material parameters will be intensified. The several kinds of hardening (isotropic, kinematic) cannot be distinguished exactly in this case.

c) The less quantities were taken into account in the vector of the material parameters to be identified the better were the results of the optimization process. Especially the consideration of parameters σ_{F_0} and β caused an increasing instability of the optimization results. Therefore it is recommendable to determine the initial yield stress analyzing simple experiments (e. g. with homogeneous stress-strain states) and to keep the auxiliary parameter β (which avoids an infinite derivative of the hardening rule (6) with respect to the equivalent plastic strain) on a given initial value during the optimization process.

d) In general, the cost function (8) may have various local minima. The choice of the initial values of the parameters to be optimized is decisive for the determination of the minimum to be found. It need not be the global optimum because the gradient based methods always lead in the direction of a descent to be found at the beginning of the optimization process.

e) It was found that errors in the measured displacements change the structure of the cost function (8) and intensify the ill-posedness of the optimization problem. This fact has a distinct effect on the optimization results and the convergence of the method.

f) The expected optimal values of the material parameters have been identified nearly regardless of the choice of the material points in which measured and calculated displacements were compared, even if no one of these points has been plastified during the entire load history (if loading, un- and reloading were taken into account). In this case stress transfers are responsible for a non-elastic behaviour of really elastic points.

g) A significant influence on the results of the optimization could not be determined using either the Euler backward method or the Crank-Nicolson method for the integration of the deformation law and for sensitivity analysis.

5. Conclusions

In this paper an inverse method was proposed to identify material parameters for elasto-plastic deformation laws with isotropic and kinematic hardening analyzing inhomogeneous two-dimensional displacement fields of notched four-point bending specimens. The inverse problem was solved using nonlinear gradient based optimization methods with semianalytical sensitivity analysis. During each optimization step comparative displacement values were calculated by nonlinear Finite Element analysis. Measured values were taken from the surface of the specimens by the Moiré method. The proposed method was verified using artificially measured displacement fields generated with given material parameters and load history. The results of the optimzation process depend on several factors. In an extensive numerical study was shown that in most of the cases the expected parameters could be determined. The cost function has several local minima due to the ill-posedness of the investigated problem. For identifying acceptable material parameters a careful choice of initial information is necessary (applicable deformation laws, accurate measurements, careful estimation of the initial values of the material parameters to be optimized, etc.). The proposed method is successfully applied for the determination of parameters of X 6 Cr Ni Ti 18 10 steel analyzing real experiments on specimens produced from a sheet metal considering the initial orthotropy of the material.

References

1. Eringen, A.C.: *Mechanics of Continua*, Krieger, Huntington New York, 1980.
2. Hadamard, J.: *Lectures on the Cauchy problem in linear partial differential equations*, Yale University Press, New Haven, 1923.
3. Kreißig, R.: Parameteridentifikation inelastischer Stoffgesetze, *Technische Mechanik* **Band 16** (1996), 97-106.
4. Bohnsack, E.: Zur Identifikation von Verfestigungsparametern elastisch-plastischer Stoffgesetze an inhomogen verformten Proben, Diss., TU Chemnitz-Zwickau, 1997.
5. Michael, D., Meyer, A.: Some remarks on the simulation of elasto-plastic problems on parallel computers, Preprint-Reihe der Chemnitzer DFG-Forschergruppe "Scientific Parallel Computing" SPC95-13, Fakultät für Mathematik der TU Chemnitz-Zwickau, 1995.
6. Mahnken, R., Stein, E.: Identification of Parameters for visco-plastic Models via Finite-Element Methods and Gradient Methods, IBNM-Bericht 93/5, Institut für Baumechanik und Numerische Mechanik der Universität Hannover, 1995.
7. Gelin, J.-C., Ghouati, O.: An inverse method for material parameters estimation in the inelastic range, *Computational Mechanics* **16** (1995), 143-150.
8. Kreißig, R.: *Einführung in die Plastizitätstheorie*, Fachbuchverlag, Leipzig, 1992.
9. Dennis J.E., Schnabel, R.B.: *Numerical Methods for Unconstrained Optimization and Nonlinear Equations*, Prentice-Hall Inc., Englewood Cliffs, New Jersey, 1983.
10. Görke, U.-J., Kreißig, R.: Einflußfaktoren bei der Identifikation von Materialparametern elastisch-plastischer Deformationsgesetze aus inhomogenen Verschiebungsfeldern. Preprint-Reihe des Chemnitzer SFB 393 (SFB393/97-04), Institut für Mechanik der TU Chemnitz-Zwickau, 1997.

DETERMINATION OF PARAMETERS IN ELASTO-PLASTIC MODELS OF ALUMINIUM

M.H.H. MEUWISSEN, C.W.J. OOMENS, F.P.T. BAAIJENS,
R. PETTERSON AND J.D. JANSSEN
*Department of Mechanical Engineering,
Eindhoven University of Technology,
P.O.Box 513, 5600 MB Eindhoven, The Netherlands*

Abstract. A mixed numerical-experimental method is used to determine parameters in elasto-plastic constitutive models. An aluminium plate of non-standard geometry is mounted in a uniaxial tensile testing machine at which some adjustments are made to carry out shear tests. The sample is loaded and the total clamp force in the shearing direction is recorded. Furthermore, the displacements of material points on the surface of the sample are measured optically. These measurements are used to determine parameters in two Von Mises plasticity models. The first model incorporates a linear hardening law and the second one a Nadai hardening law. The models are evaluated by examining the estimated parameter values and the differences between measured and calculated displacements and forces.

1. Introduction

To characterise the mechanical behaviour of a material quantitatively, a constitutive model must be chosen and the parameters in this model must be determined from measurements. In standard methods, the determination of parameters is based on the use of test samples with a well defined standardised geometry and loading, such that certain conditions on the stress and strain fields are satisfied in (a part of) the sample. Then, the unknown parameters are obtained via analytical relations from measurements of, for example, forces, torques, displacements and/or twist angles.

In some situations, the application of standard methods is inconvenient or even impossible. Examples can be found in many metal forming processes where materials are loaded at high strains, high strain rates, possibly

H. Sol and C.W.J. Oomens (eds.), Material Identification Using Mixed Numerical Experimental Methods, 71-80.

combined with high temperatures. Commonly used tests in this area are uniaxial tensile and compression tests, torsion tests, and bending tests. If these tests are to be carried out in a domain comparable to the actual forming process (*i.e.* at high strains, high strain rates and high temperatures), they fail due to disturbances like necking, fracture, friction between die and sample and/or inhomogeneous deformation heating.

For these reasons, an increasing interest is found in literature in so-called mixed numerical-experimental methods based on the use of non-standard experiments in which the conditions on stress and strain fields are relaxed [1–5]. In many of these applications, the sample geometry and/or loading is still close to those of standard methods. Moreover, the experimental information is limited to quantities measured at the boundaries of sample(s).

In the present study, the mixed numerical-experimental method as proposed by Hendriks [6] is applied to determine parameters in elasto-plastic models. In this method the geometry and/or loading of samples is non-standard. In addition, next to boundary information, field information is measured over the sample's surface. This extra information is expected to be valuable in parameter determination and model evaluation, especially for inhomogeneous and/or anisotropic material behaviour.

The current research is a pilot for the application of the mixed numerical-experimental method as proposed by Hendriks for parameter determination at high strains and high strain rates. For these initial studies the experiment is still quasi-static and the strain levels are low, but the geometry of the sample is non-standard.

2. The mixed numerical-experimental method

The mixed numerical-experimental method used to determine constitutive parameters, is characterised by the following three components:

- One or more experiments in which, next to boundary information, non-uniform displacement fields are measured on samples that are more or less arbitrarily shaped and possibly multi-axially loaded.
- A numerical (finite element) analysis of the experiment(s) in which an assumed constitutive model of the material behaviour is incorporated.
- An estimation algorithm which determines estimates of the unknown constitutive parameters iteratively based on initial guesses and the difference between measured and calculated output.

New estimates of the unknown parameters are determined by minimising the following sum of squares:

$$J = (\underline{m} - \underline{h}(\underline{\theta}))^T \underline{V} (\underline{m} - \underline{h}(\underline{\theta})) + (\underline{\theta}_0 - \underline{\theta})^T \underline{W} (\underline{\theta}_0 - \underline{\theta}) \qquad (1)$$

where $\underline{m}$ is a column containing the measured output (for example displacements and/or forces), $\underline{h}$ the corresponding output of the finite element analysis, $\underline{\theta}$ the *a priori* unknown parameters, $\underline{\theta}_0$ the initial guess of the parameters and $\underline{V}$ and $\underline{W}$ two positive definite symmetric weighting matrices. In the present study, the latter two matrices are chosen to be diagonal. The sum of squares is minimised iteratively using the Gauss-Newton scheme. Iteration is continued until one of the following criteria is satisfied: the sum of squared residual output is smaller than a critical value or the sum of squared parameter adjustments is smaller than a critical value. More details about the algorithm can be found in Ref. [6].

3. Experimental details

3.1. SAMPLE AND EXPERIMENTAL SETUP

The investigated material is a commercially available aluminium alloy 1200. From a 2.0 mm thick plate, a sample with dimensions 40 × 80 mm is machined (see Fig. 1).

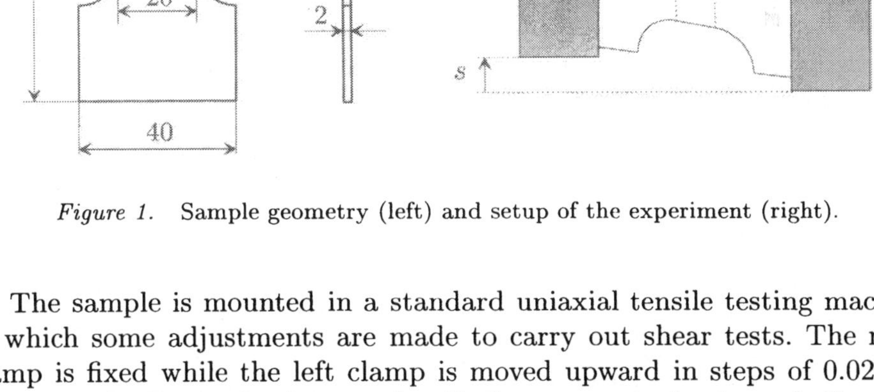

Figure 1. Sample geometry (left) and setup of the experiment (right).

The sample is mounted in a standard uniaxial tensile testing machine at which some adjustments are made to carry out shear tests. The right clamp is fixed while the left clamp is moved upward in steps of 0.02 mm (see Fig. 1). For both clamps, the displacements in horizontal direction are suppressed. Due to the geometry of the sample, the highest strains are found in the central region.

The recorded clamp force F versus load step number is shown in Fig. 2. At each step marked by '*', the displacements of points distributed over the sample's surface are measured optically using the digital image correlation technique described in the next section.

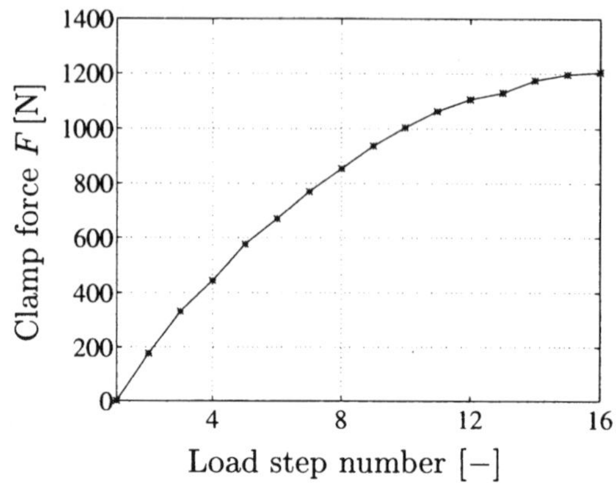

Figure 2. Measured clamp force F versus load step number. At steps denoted by '*', the clamp force and clamp displacement are measured. Load step 16 corresponds to a clamp displacement of $s = 0.3\,\text{mm}$.

3.2. METHOD OF DISPLACEMENT MEASUREMENT

The digital image correlation technique applied here is originally developed at the University of South Carolina [7]. In this technique, two digitised video images of a sample are compared in order to determine displacements and displacement gradients caused by deformation in the sample.

The experimental setup used to acquire the video images is schematically shown in Fig. 3. Images of the sample in reference and deformed configuration are obtained with a CCD camera (Charge Coupled Device, PULNIX TM-765) and a framegrabber (MATROX MAGIC) controlled by an IBM-compatible PC. The images are stored as 768×512 sets of 8-bit numbers. Each number in this set represents the intensity of the light impinging on a particular pixel of the CCD camera. A value of 0 corresponds to black and a value of 255 to white, while the intermediate values correspond to different shades of gray.

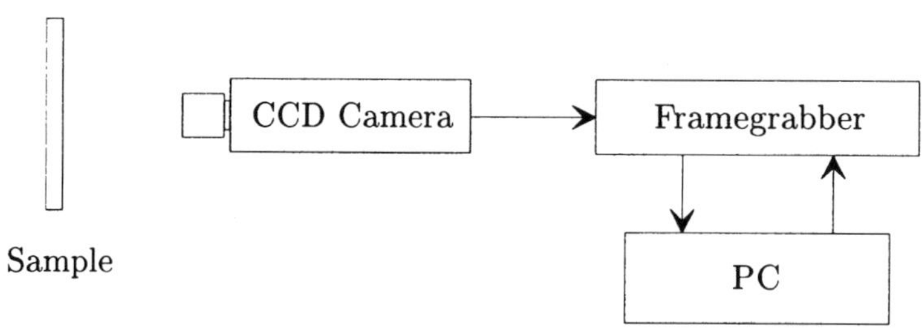

Figure 3. Experimental setup for displacement field measurement.

Let (x_D, y_D) denote the *a priori* unknown position of a point in the second image (*i.e.* the image of the sample in the deformed configuration) which was at position (x_R, y_R) in the first image (*i.e.* the image of the sample in the reference configuration). Furthermore, let $b_1(x,y)$ denote the gray value at position (x,y) in the first image and $b_2(x,y)$ the gray value at position (x,y) in the second image. Gray values of the digitised images are only available at discrete positions (pixels). The gray values at intermediate positions are approximated using bicubic spline interpolation. Now, a subset $a_1(x,y)$ of $b_1(x,y)$ is defined around the point (x_R, y_R). After deformation, subset $a_1(x,y)$ is mapped onto subset $a_2(x^*, y^*)$ of $b_2(x,y)$ around (x_D, y_D). It is assumed that the deformation of the sample is parallel to the image plane. The deformation of the subset $a_1(x,y)$ is described by the first terms of a Taylor series expansion:

$$\begin{aligned} x^* &\approx x + u + \tfrac{\partial u}{\partial x}\Delta x + \tfrac{\partial u}{\partial y}\Delta y \\ y^* &\approx y + v + \tfrac{\partial v}{\partial x}\Delta x + \tfrac{\partial v}{\partial y}\Delta y \end{aligned} \tag{2}$$

where u and v are the displacements of point (x_R, y_R) in x- and y-direction respectively, $\frac{\partial u}{\partial x}, \frac{\partial u}{\partial y}, \frac{\partial v}{\partial x}, \frac{\partial v}{\partial y}$ are the displacements gradients at point (x_R, y_R), and Δx and Δy are given by:

$$\begin{aligned} \Delta x &= x - x_R \\ \Delta y &= y - y_R \end{aligned} \tag{3}$$

The unknown parameters u, v, $\frac{\partial u}{\partial x}, \frac{\partial u}{\partial y}, \frac{\partial v}{\partial x}, \frac{\partial v}{\partial y}$ in Eq. (2) are determined by maximising the following correlation function r:

$$r(u, v, \frac{\partial u}{\partial x}, \frac{\partial u}{\partial y}, \frac{\partial v}{\partial x}, \frac{\partial v}{\partial y}) = \frac{1}{C} \int\!\!\int_{\Delta A} a_1(x,y)\, a_2(x^*, y^*)\, dx\, dy \tag{4}$$

where ΔA is the area of the subset $a_1(x,y)$ and C is a scaling factor:

$$C = \sqrt{\int\!\!\int_{\Delta A} a_1(x,y)^2\, dx\, dy \int\!\!\int_{\Delta A} a_2(x^*, y^*)^2\, dx\, dy} \tag{5}$$

For a perfect match of the subsets the correlation function r equals 1.

3.3. MEASURED DISPLACEMENT FIELDS

The results of the digital image correlation technique are summarised in Fig. 4. This figure shows the displacement paths of 114 points on the surface of the sample during the shear test. The original positions of the points are marked by plus signs. The maximal displacements at the final load step are approximately 0.25 mm near the left clamp which corresponds to

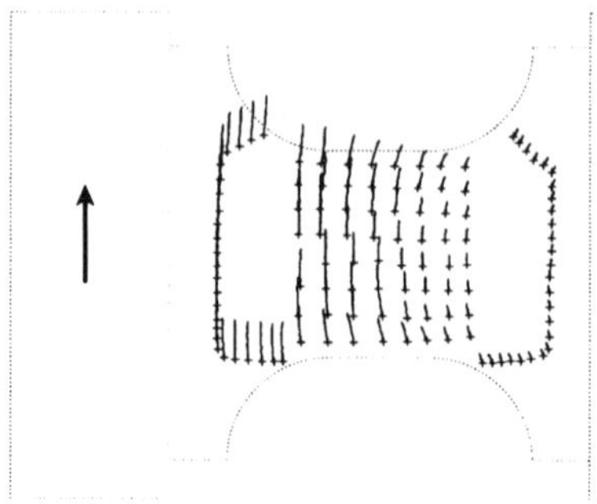

Figure 4. Displacement paths of 114 points over load steps 2–16. Initial positions of the points are marked by plus signs.

approximately 5 pixels. The maximal equivalent strains at the final load step are less than 1%.

4. Setup of estimation process

4.1. FINITE ELEMENT MODEL AND BOUNDARY CONDITIONS

A finite element model of the central region of the sample is implemented in the finite element code MARC [8] using 1702 4-noded isoparametric plane stress elements (see Fig. 5).

The positions of material points at the border of this region are used to define the left and right borders of the finite element model. The measured displacements of these points are used as kinematic boundary conditions in the finite element model. The boundary displacements are applied in 45 increments.

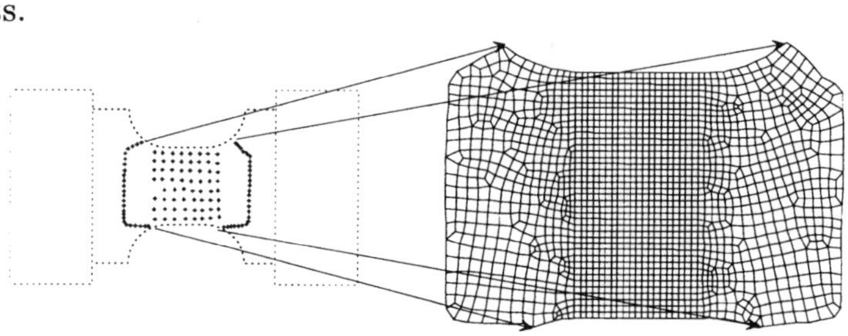

Figure 5. Generation of finite element mesh of a part of the sample.

4.2. CONSTITUTIVE MODELS

To describe the mechanical behaviour of the aluminium, an elasto-plastic constitutive model is adopted. In the elastic region, the material is assumed to behave according to Hooke's law. In the plastic region, two different models are adopted. In both models, the Von Mises yield criterion with an associated flow law is assumed. The models differ in the type of work hardening law. The first model incorporates a linear work hardening law:

$$\sigma_v = \sigma_{v0} + L\varepsilon_p \tag{6}$$

where σ_v is the current yield stress, σ_{v0} the initial yield stress, ε_p the equivalent plastic strain and L the plastic modulus. The second model incorporates a Nadai work hardening law:

$$\sigma_v = \sigma_{v0} + c\varepsilon_p^n \tag{7}$$

where c and n are two parameters.

4.3. PARAMETERS $\underline{\theta}$, MEASUREMENTS $\underline{m}$ AND WEIGHTING MATRICES

The resolution of the measurement system was not high enough to measure the small elastic strains sufficiently accurately. Therefore, the Young's modulus E and Poisson's ratio ν were taken from uniaxial tensile tests ($E = 7.0 \cdot 10^{+4}$ N/mm^2, $\nu = 0.31$). In the first model, the following unknown parameters are estimated:

$$\underline{\theta}_1^T = [\sigma_{v0}, L] \tag{8}$$

The initial estimates of these parameters are:

$$\underline{\theta}_{10}^T = [80.0, 5000.0] \tag{9}$$

For the second model, it proved to be difficult to estimate the 3 parameters in the Nadai law simultaneously. This is caused by near dependencies of the parameters during the estimation process. Therefore, the initial yield stress σ_{v0} is set to 50.0 N/mm^2. Then, the unknown parameters in the model are:

$$\underline{\theta}_2^T = [c, n] \tag{10}$$

and the initial estimates:

$$\underline{\theta}_{20}^T = [150.0, 0.10] \tag{11}$$

The diagonal elements of the weighting matrices $\underline{W}$ (see Eq. (1)) are set to small numbers so that the initial estimates of the parameters have no effect on the final estimates.

The measured clamp force F (see Fig. 1) and the measured displacements of the 63 points in the central part of the sample (see Fig. 4) are

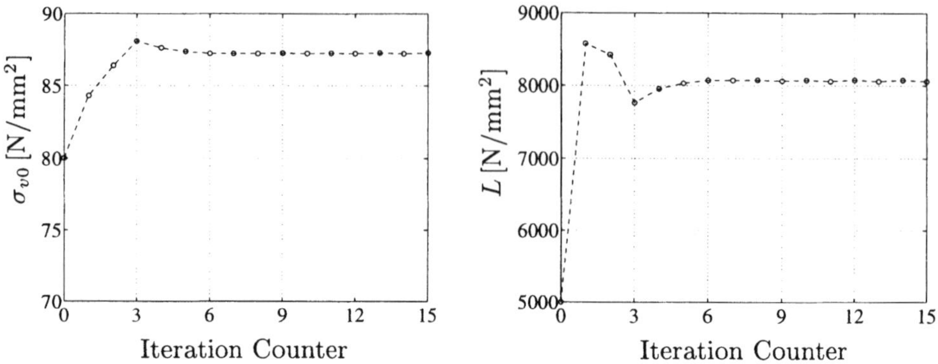

Figure 6. Estimated initial yield stress σ_{v0} (left) and plastic modulus L (right) versus iteration counter for the first model.

used to determine the unknown parameters. For all steps marked by '*' in Fig. 2, these measurements are stored in the column $\underline{m}$ (see Eq. (1)). The column $\underline{m}$ contains a total of 1890 displacement components and 15 force components. The diagonal elements of the weighting matrix $\underline{V}$ corresponding to displacement components are set to 10^0 while the diagonal elements corresponding to force components are set to 10^{-5}. These choices are based on engineering judgement.

5. Estimation results

Fig. 6 shows the estimated parameters versus iteration counter for the first model. Within 6 iterations, the estimates converge to stable values, which are physically reasonable. Several estimation runs were performed starting with different initial parameter estimates. In most cases, convergence was established to the same final estimates. In some cases (very bad initial estimates) the parameters did not converge to stable values.

The residuals (*i.e.* the difference between measured and predicted output) are shown in Fig. 7. The average magnitude of the residuals is at most 10% of the average magnitude of measurements. Hence, the model gives a good prediction of the observed behaviour. However, deterministic patterns are still present in both the residual forces and displacements, which indicates the presence of one or more (small) modelling error(s).

Fig. 8 shows the estimated parameters versus iteration counter for the second model. The estimates converge to stable values after 12 iterations.

The residuals are shown in Fig. 9. For this model the residuals are slightly smaller, although deterministic patterns are still visible. Performing estimation runs with different values of the initial yield stress σ_{v0} leads to almost identical fits of the measurements.

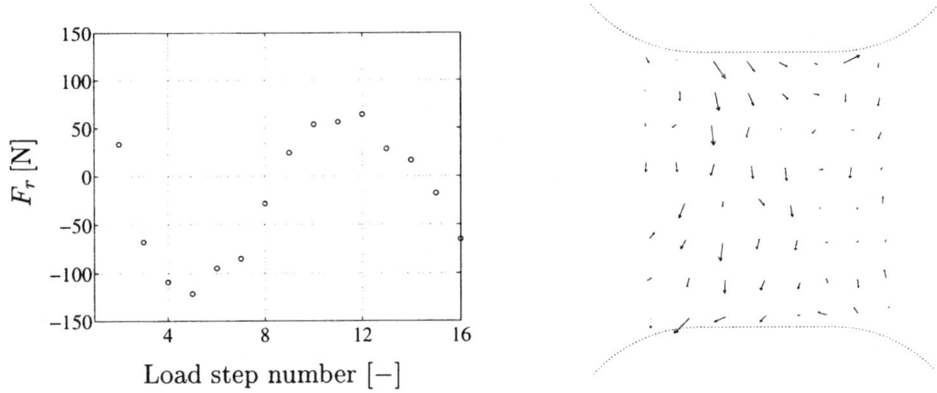

Figure 7. Residual clamp force F_r (left) and residual displacements at load step 16 (right). Estimated standard deviation of the residual displacements: $\hat{s} \approx 5.6 \cdot 10^{-3}$ mm. Average measured displacements at this step are of order 10^{-1} mm.

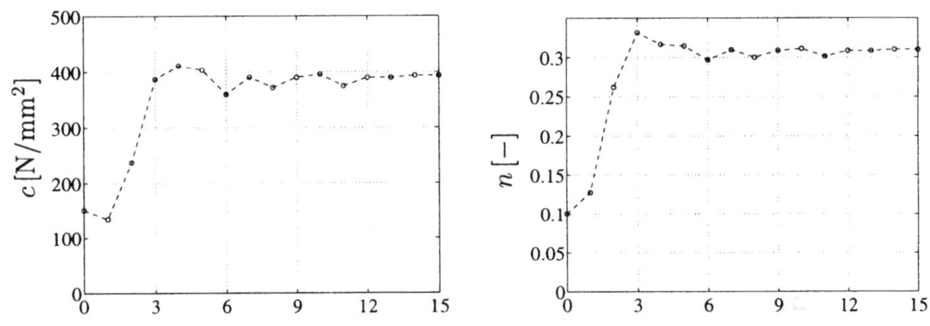

Figure 8. Estimated parameters in the Nadai hardening law: c (left) and n (right). The initial yield stress σ_{v0} is fixed to $50.0\,\mathrm{N/mm^2}$.

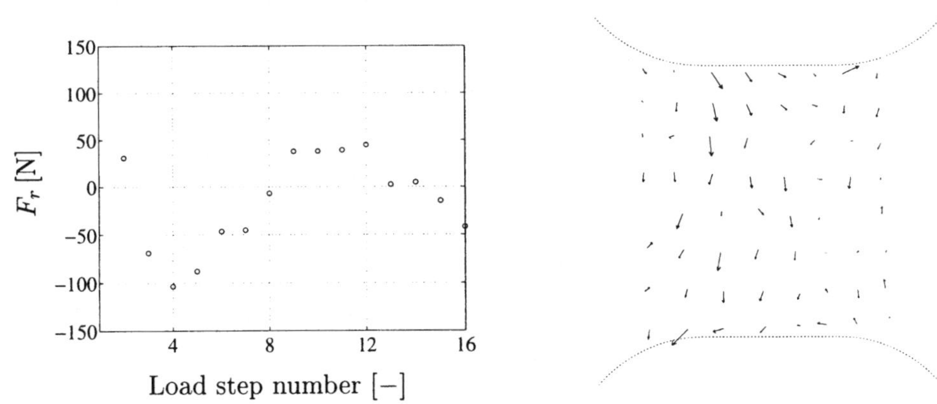

Figure 9. Residual clamp force F_r (left) and residual displacements at load step 16 (right). Estimated standard deviation of the residual displacements: $\hat{s} \approx 5.3 \cdot 10^{-3}$ mm. Average measured displacements at this step are of order 10^{-1} mm.

6. Discussion

A mixed numerical-experimental method is used to determine parameters in two different elasto-plastic constitutive models incorporating a linear and Nadai work hardening law. Both models gave accurate predictions of the observed clamp forces and displacement fields. The prediction of model incorporating the Nadai hardening law is slightly better.

Nevertheless, deterministic patterns are visible in the residuals which indicate the presence of one or more (small) modelling error(s). These patterns are probably caused by measurement errors in the displacements of the points used as boundary conditions in the model and/or by defects in the constitutive equations.

In future research, the behaviour of the material at higher strains and cyclic loading will be addressed.

Acknowledgements

This work is part of a research project financially supported by the Dutch Technology Foundation (STW).

References

1. Cailletaud, G., Pilvin, P.: Identification and inverse problems related to material behaviour, in: *Inverse problems in engineering mechanics* (eds. H.D. Bui & M. Tanaka *et al.*), Balkema, Rotterdam, 79–86, 1994.
2. Gavrus, A., Massoni, E., Chenot, J.L.: Computer aided rheology for non linear large strain thermo-viscoplastic behaviour formulated as an inverse problem, in: *Inverse problems in engineering mechanics* (eds. H.D. Bui & M. Tanaka *et al.*), Balkema, Rotterdam, 123–130, 1994.
3. Gelin, J.C., Ghouati, O.: Inverse solution procedure for material parameters identification in large plastic deformations, *Comm. Num. Meth. Eng.* **12** (1996), 189–192.
4. Khoddam, S., Lam, Y.C., Thomson, P.F.: Numerical results achieved with an inverse computational method for determining the constitutive parameters using hot torsion test results, *Steel Research* **67** (1996), pp. 39–43.
5. Kusiak, J., Kawalla, R., Pietrzyk, M., Pircher, H.: Inverse analysis applied to the evaluation of material parameters in the history dependent flow stress equation in hot forming of metals, *J. of Materials Processing Technology* **60** (1996), 455–461.
6. Hendriks, M.A.N.: *Identification of the mechanical behavior of solid materials*, Ph.D. thesis, Eindhoven University of Technology, The Netherlands, 1991.
7. Bruck, H.A., McNiell, S.R., Sutton, M.A., Peters III, W.H.: Digital image correlation using Newton-Raphson method of partial differential correction, *Experimental Mechanics* **29** (1989), 261–267.
8. MARC Analysis Research Corporation, 1994, Vol. A: User Information, Vol. B: Element Library, Vol. C: Program Input.

MATERIAL PARAMETER IDENTIFICATION FOR LARGE DEFORMATION PLASTICITY MODELS

V.V. TOROPOV[1], F. YOSHIDA[2] and E. VAN DER GIESSEN[3]

[1]*Department of Civil and Environmental Engineering, University of Bradford, Bradford, West Yorkshire, BD7 1DP, UK*

[2]*Department of Mechanical Engineering, Hiroshima University, 1-4-1, Kagamiyama, Higashi-Hiroshima, 739, Japan*

[3]*Laboratory for Engineering Mechanics, Delft University of Technology, Mekelweg 2, NL-2628 CD Delft, The Netherlands*

1. Introduction

In order to numerically simulate manufacturing processes of structural components and their mechanical behaviour in service with sufficient accuracy and computational efficiency, it is necessary to develop appropriate constitutive models. All such models contain a number of material parameters (phenomenological coefficients) which are to be determined from some kind of experiments with a material specimen. Traditionally, the determination of material properties of engineering materials is carried out when the strain and stress fields inside a specimen are to a high degree uniform (e.g. the tensile test) and material parameters are derived from the test data analytically.

There are many cases when the application of this traditional approach is very difficult and sometimes impossible. The main reasons for that can be:

- the stress state in a specimen is (or becomes) inhomogeneous (e.g. during large deformation plasticity involving necking);
- the material may exhibit anisotropic behaviour.

To overcome these difficulties, Toropov and Van der Giessen (1993) proposed to carry out *nontrivial experiments*, i.e. experiments where the specimen used for the parameter assessment involves inhomogeneous and multiaxial stress and strain fields. It allows to use the whole structural elements as specimens. The values of material parameters cannot then be derived from the experiment immediately; in contrast, a numerical analysis is necessary to simulate the actual experiment; however, this requires that the material parameters are known. The identification problem can then be formulated as an optimization problem where the function to be minimized is some error function that expresses the difference between numerical simulation results and experimental data.

H. Sol and C.W.J. Oomens (eds.), Material Identification Using Mixed Numerical Experimental Methods, 81-92.

2. Formulation of the Material Parameter Identification Problem as an Optimization Problem

Let us consider the phenomenological coefficients of the constitutive model to be identified as components of the vector $x \in R^N$. Then the optimization problem can be formulated as follows:

Find the vector x that minimizes the objective function

$$F(x) = \sum_{\alpha=1}^{M} \theta^{\alpha} F^{\alpha}(x), \quad A_i \leq x_i \leq B_i \quad (i = 1, \ldots, N) \tag{1}$$

where M is the total number of individual specific response quantities (denoted by α, which can be measured in the course of experiments and then obtained as a result of the numerical simulation; $F^{\alpha}(x)$ is the dimensionless function

$$F^{\alpha}(x) = \{ \sum_{s=1}^{S_\alpha} [R_s^{\alpha} - R^{\alpha}(x, \tau_s^{\alpha})]^2 \} / \{ \sum_{s=1}^{S_\alpha} [R_s^{\alpha}]^2 \}, \tag{2}$$

which measures the deviation between the computed α -th individual response and the observed one from the experiment; τ^{α} is a parameter which defines the history of the process in the course of the experiment (e.g. the time or the loading parameter), and the values τ_s^{α} ($\alpha = 1, \ldots, M$, $s = 1, \ldots, S_\alpha$) define the discrete set of S_α data points; R_s^{α} is the value of the α-th measured response quantity corresponding to the value of the experiment history parameter τ_s^{α}; $R^{\alpha}(\mathbf{x}, \tau_s^{\alpha})$ is the value of the same response quantity obtained from the numerical simulation; θ^{α} is the weight coefficient which determines the relative contribution of information yielded by the α-th set of experimental data; A_i , B_i are lower and upper limits on the values of material parameters stipulated by physical considerations.

3. Multipoint Approximation Technique

The optimization problem (1) has the following characteristic features: (*i*) the objective function is an implicit function of parameters x, (*ii*) to calculate values of this function for the specific set of parameters x means to use a nonlinear numerical (e.g. finite element) simulation of the process under consideration, which usually involves a large amount of computer time, (*iii*) function values present some level of noise, i.e. can only be estimated with finite accuracy.

The direct implementation of any conventional nonlinear mathematical programming techniques would involve too large amount of computer time and, moreover, the convergence of a method cannot be guaranteed due to the presence of numerically induced noise in the objective function values and (or) its derivatives. The iterative multipoint approximation concept (Toropov *et al.*, 1993) has been developed to deal with these complications, and has been successfully used to solve various structural optimizations and identification problems.

The technique is based on the iterative approximation of computationally expensive and noisy functions $F^{\alpha}(\boldsymbol{x})$, ($\alpha = 1,\dots, M$) by simplified noiseless functions. The initial optimization problem (1) is replaced then with the succession of simpler mathematical programming subproblems as follows:

Find the vector $\boldsymbol{x}_k^*$ that minimizes the objective function

$$\tilde{F}_k(\boldsymbol{x}) = \sum_{\alpha=1}^{M} \theta^{\alpha} \tilde{F}_k^{\alpha}(\boldsymbol{x}), \quad A_i^k \le x_i \le B_i^k,\ A_i^k \ge A_i,\ B_i^k \le B_i\ (i = 1,\dots,N) \tag{3}$$

where k is the iteration number. The move limits A_i^k and B_i^k define a subregion of the optimization parameter space where the simplified functions $\tilde{F}_k^{\alpha}(\boldsymbol{x})$ are considered as current approximations of the original implicit functions $F^{\alpha}(\boldsymbol{x})$. To estimate their accuracy, the error parameter $r_k = \left|[F(\boldsymbol{x}_k^*) - \tilde{F}_k(\boldsymbol{x}_k^*)]/F(\boldsymbol{x}_k^*)\right|$ is evaluated. It gives a measure of the discrepancy between the values of initial functions and the simplified ones at a point corresponding to the solution of the current optimization subproblem. Any of conventional optimization techniques can be used to solve a subproblem (3) because the functions, involved in its formulation, are simple and noiseless.

To construct the simplified expression for the $\tilde{F}_k^{\alpha}(\boldsymbol{x})$ in (3), the methods of regression analysis are implemented. They are intended for obtaining an analytical expression that reflects the behaviour of an object considered as a function of its parameters, based on a discrete set of experimental results. *An experiment* means here and in the remainder of this section *a computational experiment* using the numerical model of the process under consideration. It is essential to note that we do not intend to construct simplified expressions that are adequate in the whole of the search region determined by side constraints A_i and B_i in (1) because it takes too large number of numerical experiments in case of a real-life multiparameter problem. Therefore, it is proposed to construct such expressions iteratively only for separate search subregions determined by move limits at each step of the iterative process. Thus, the simplified functions $\tilde{F}_k^{\alpha}(\boldsymbol{x})$ give piece-wise approximations of the initial functions $F^{\alpha}(\boldsymbol{x})$.

Let us now consider the problem of formulating the simplified functions $\tilde{F}_k^{\alpha}(\boldsymbol{x})$. It is assumed that a simplified function can be expressed in the following general form:

$$\tilde{F}_k^{\alpha}(\boldsymbol{x}) = \tilde{F}_k^{\alpha}(\boldsymbol{x}, \boldsymbol{a}^{\alpha}). \tag{4}$$

The vector $\boldsymbol{a}^{\alpha} = (a_1^{\alpha},\dots,a_L^{\alpha})$ in (4) consists of so-called *tuning parameters*, i.e. free parameters the value of which are determined on the basis of numerical experiments at points located in the optimization variable space R^N in accordance with some design (plan) of experiments. The tuning parameters are found using the weighted least-squares method as the solution of the following minimization problem:

$$\min G_k^{\alpha}(\boldsymbol{a}^{\alpha}), \quad G_k^{\alpha}(\boldsymbol{a}^{\alpha}) = \sum_{p=1}^{P} w_p^{\alpha} [F^{\alpha}(\boldsymbol{x}_p) - \tilde{F}_k^{\alpha}(\boldsymbol{x}_p, \boldsymbol{a}^{\alpha})]^2 \tag{5}$$

where P is the total number of points in the plan of experiments; $\boldsymbol{x}_p$ is the vector of optimization variables that defines the current point; w_p^α is the weight coefficient that characterizes the relative contribution of the p-th experiment's information.

The solution of the optimization problem (5) is the vector $\boldsymbol{a}^\alpha$ that makes up the simplified function (4). A proper choice of the weight coefficients w_p^α ($p = 1,..., P$) can substantially improve the properties of simplified functions $\tilde{F}_k^\alpha(\boldsymbol{x}, \boldsymbol{a}^\alpha)$, see Toropov and Van der Giessen (1993) for more details.

After formulation of the functions $\tilde{F}_k^\alpha(\boldsymbol{x}, \boldsymbol{a}^\alpha)$, the current optimization problem (3) is solved and the error parameter r_k for the obtained point $\boldsymbol{x}_k^*$ is estimated. Next, the move limits A_i^k and B_i^k are to be determined the for the next iteration. First, the condition is checked of whether the error parameter r_k was sufficiently small, i.e. the simplified models of a current iteration were accurate enough. If it is not satisfied, then the size of the search subregion of the next step must be reduced. If the simplified models were accurate (r_k was sufficiently small), the decision is to be made on how to move the search subregion. If the point obtained, $\boldsymbol{x}_k^*$, is located inside the k-th search subregion (none of the move limits is active), then that point can be considered as the current approximation of the solution $\boldsymbol{x}^*$. In that case, the next search subregion should be reduced and the other conditions of the search termination should be checked. Otherwise, if some of the side constraints in (3) satisfied as equalities, the search must be continued. This means that the search subregion must be moved (keeping its size) further in the direction $(\boldsymbol{x}_k^* - \boldsymbol{x}_{k-1}^*)$ of the previous step. Depending on the accuracy of approximations, either a new plan of experiments in the next search subregion should be chosen or the approximations can be used once again in the new search subregion. The search process is terminated when (*i*) the error parameter r_k is sufficiently small, (*ii*) none of move limits is active and (*iii*) the subregion has reached a required small size.

Let us now consider the problem of formulating the simplified expressions $\tilde{F}_k^\alpha(\boldsymbol{x}, \boldsymbol{a}^\alpha)$. The efficiency of the optimization technique depends greatly on their accuracy. Note that properly chosen simplified expressions in our case of the optimization problem (1) with no behavioural constraints imposed, must allow for internal minimum point inside the search region defined by side constraints A_i and B_i, for otherwise the convergence of the method can be slow; i.e., the number of calls for the evaluation of the functions $F^\alpha(\boldsymbol{x})$ would be large. The simplest form of an expression which satisfies the above requirement, is a full quadratic polynomial in $\boldsymbol{x}$ (linear in $\boldsymbol{a}$). If there are no other considerations available, then it can be considered as a reasonable choice. However, it requires at least $N\ (N+1)/2 + N + 1$ calls for the evaluation of the functions $F^\alpha(\boldsymbol{x})$ in (2), which can be a large number in the case of a multiparameter identification problem where N can easily be of the order of 10.

There is an alternative approach based on so-called *mechanistic models*, which finds increasing application in empirical model-building (Box and Draper, 1987). The parameter estimation of such models requires the implementation of the most general (nonlinear) form of the least-squares method. These models are constructed on the basis of physical considerations, which can sometimes provide clues to the nature of a phenomena under consideration. The designer of such a mechanistic model can typically use *a priori* information, such as analytical solutions for simplified geometrical shape, loading or boundary conditions of the specimen under consideration. Clearly, in this approach, the researcher's experience and engineering judgement is essential to create high-quality approximations. Typically, the available information presents the description of the process under consideration as a function of the experiment history parameter τ^α. Such information can be used for the formulation of the simplified model in the following form:

$$\tilde{F}_k^\alpha(\boldsymbol{x}, \boldsymbol{a}^\alpha) = \{ \sum_{s=1}^{S_\alpha} [R_s^\alpha - \tilde{R}_k^\alpha(\boldsymbol{x}, \boldsymbol{a}^\alpha, \tau_s^\alpha)]^2 \} / \{ \sum_{s=1}^{S_\alpha} [R_s^\alpha]^2 \}, \qquad (6)$$

where $\tilde{R}_k^\alpha(\boldsymbol{x}, \boldsymbol{a}^\alpha, \tau^\alpha)$ is a simplified model of the process under consideration.

This approach has been successfully used by Toropov and Van der Giessen (1993) in the problem of material parameter identification for a large deformation plasticity model from the experimental data obtained using a solid bar in torsion. In order to construct a mechanistic model (i.e. physically motivated high quality approximations), analytical and simplified numerical solutions for a tubular specimen have been used. As a drawback of that approach, its lack of generality can be mentioned because every new problem involves a considerable amount of analytical work to obtain high quality mechanistic approximations. A more general approach can be beneficial in cases when the numerical noise affects mostly the accuracy of derivatives of the error function (1) and to a lesser extent the function values. Then it can be easier (and more straightforward) to use an algorithm of sequential quadratic programming (SQP) which builds up an approximation of the inverse of the Hessian matrix using the objective function values and its derivatives. Again, the multipoint approximations of the objective functions should be used iteratively instead of original function values. It should be noted that in problems of parameter identification for nonlinear constitutive models the traditional use of finite differences for evaluation of derivatives would almost certainly spoil the convergence of the optimizer because the accuracy of derivatives is severely affected by the noise. As the simplified functions are used to evaluate the derivatives only, the requirement for simplified functions to allow for internal minimum can be waived. A similar approach was used by Marti (1991) who replaced the gradients of stochastic functions by gradients of the polynomial response surface models. As the approximations are constructed in each iteration to evaluate the derivatives only, a fairly simple approximation types can be used, e.g. quadratic polynomials without cross-terms:

$$\tilde{F}_k^\alpha(\boldsymbol{x}, \boldsymbol{a}^\alpha) = a_0^\alpha + \sum_{i=1}^{N} (a_{2(i-1)+1}^\alpha x_i + a_{2(i-1)+2}^\alpha x_i^2) \qquad (7)$$

which contains 2N+1 tuning parameters to be found in each iteration by the least-squares surface fitting.

4. Application to Models for Large Elastoplastic Deformations

As an example of the above procedure, we shall discuss the identification of material parameters in a class of constitutive models for large elastoplastic deformations. We will focus on phenomenological models that include deformation-induced anisotropy, as it occurs in engineering metals under large strains. Such models find potential application in, for instance, the simulation of industrial forming processes. The inherent physical nonlinearity of such models is further complicated by the complex history of the deformation-induced anisotropy. The simplest possible experiment - uniaxial tension - poses two major difficulties for the parameter identification in large deformation plasticity models:

- for the models that include deformation-induced anisotropy, the uniaxial response need not depend on all material parameters;
- at large strains, necking intervenes and leads to a complicated three-dimensional stress and strain state.

Then the experiment used for the material parameter identification shall be free from such instabilities and provide a means to identify all pertinent parameters for the considered class of models. Therefore, the experimental data are to be provided by the series of non-trivial experiments, i.e. experiments where the specimen used for model parameter assessment involves inhomogeneous and multiaxial stress and strain fields. The parameter identification from such tests will then require the numerical simulation to determine stresses and strains, most typically, a finite element simulation incorporating the constitutive model to be identified.

Two examples, demonstrating the application of the above technique to identification of material parameters for large deformation plasticity models, are given below.

4.1. SOLID BAR IN TORSION

The torsion test employs a solid circular bar of 150 mm length and an initial radius of 7.5 mm. The twist angle φ is measured over a gauge section of initial length of 100 mm which is believed to be sufficiently far from the ends to exclude end effects. In addition to the angle of twist, the axial displacement over the same gauge section is monitored. The reason for doing that is that for large torsional strains, the specimen will elongate when it is left free in the axial direction. This phenomenon (Swift effect) is attributed to deformation-induced anisotropy due to texture. The quantitative experimental determination of Swift effect is an important source of information for the identification of material parameters in the constitutive description of the evolving anisotropy; but it should be realized that it is a second order effect. Figure 1 shows typical results of such a torsion test on 304 stainless steel in terms of the torque T and the (engineering) axial strain ε_z as a function of twist φ. The tests were carried out at room temperature and at a twist rate of 1.1×10^{-3} rad/s.

The constitutive equations used here, are based on a combined kinematic-isotropic hardening model including a plastic spin constitutive equation (Wu and Van der Giessen, 1991). This model features the gradual fading of kinematic hardening at large strains, based on the experimental observation that typically at strains larger than

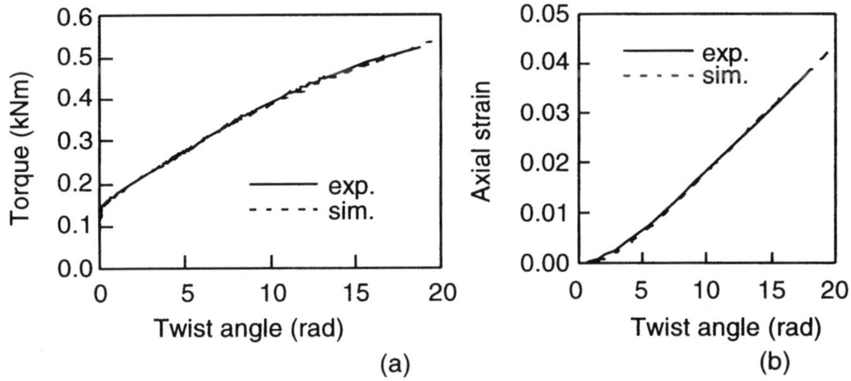

Figure 1. Simulated and measured response (torsion problem).

several ten percents no Bauschinger effect is observed. A summary of the constitutive equations is given in Table 1.

TABLE 1. Summary of constitutive equations

kinematics:	$\boldsymbol{D} = \boldsymbol{D}^e + \boldsymbol{D}^p, \quad \boldsymbol{W} = \boldsymbol{W}^e + \boldsymbol{W}^p$
(hypo-) elasticity:	$\overset{\vee}{\sigma} = \dot{\sigma} - \boldsymbol{W}^e\sigma + \sigma\boldsymbol{W}^e = \boldsymbol{L}{:}\boldsymbol{D}^e,$
	$L_{ijkl} = \frac{E}{1+\nu}[\frac{1}{2}(\delta_{ik}\delta_{jl} + \delta_{il}\delta_{jk}) + \frac{\nu}{1-2\nu}\delta_{ij}\delta_{kl}]$
yield condition:	$\varphi(\bar{s}) \equiv \mathrm{tr}\,\bar{s}^2/2 - \sigma_y^2/3, \quad \bar{s} = (s - \boldsymbol{a})/H$
hardening evolution:	$\sigma_F \equiv H\sigma_y = (1-b)\sigma_y + b\sigma_e, \quad \sigma_e = \sqrt{3\,\mathrm{tr}\boldsymbol{s}^2/2},$
	$\dot{b} = (1-b)\dot{\varepsilon}_p/\varepsilon_b, \quad \dot{\varepsilon}_p = \sqrt{2\,\mathrm{tr}\boldsymbol{D}^{p^2}/3}$
flow rule:	$\boldsymbol{D}^p = \lambda\frac{\partial\varphi}{\partial\bar{s}} = \lambda\bar{s}, \quad \boldsymbol{W}^p = \lambda\frac{1}{2}\rho(\boldsymbol{a}\bar{s} - \bar{s}\boldsymbol{a})$
back stress evolution:	$\overset{\vee}{\boldsymbol{a}} \equiv \dot{\boldsymbol{a}} - \boldsymbol{W}^e\boldsymbol{a} + \boldsymbol{a}\boldsymbol{W}^e = \mu\boldsymbol{D}^p$

Here, $\boldsymbol{D}$ and $\boldsymbol{W}$ are generic for strain-rate tensors and spin tensors, respectively, tr denotes the trace and **:** is the double dot product ($(\boldsymbol{L}{:}\boldsymbol{D})_{ij} = L_{ijkl} D_{kl}$). Further, σ and **s** are the Cauchy stress tensor and its deviator, $\boldsymbol{a}$ is the back stress tensor which specifies the translation of the yield surface's centre in σ- space and H specifies the isotropic expansion. The relative contributions of kinematic and isotropic hardening to strain hardening are governed by the parameter b ranging from 0 (pure kinematic hardening) to 1 (pure isotropic behaviour). The assumed model incorporates the following material parameters: elastic constants E and ν ; the initial von Mises yield stress σ_y; parameters $\Delta\sigma_y^{(i)}, \varepsilon_o^{(i)}$ in the expression

$$\sigma_e = \sigma_y + \sum_{i=1}^{K} \Delta\sigma_y^{(i)}[1 - \exp(-\varepsilon_p/\varepsilon_o^{(i)})] \tag{8}$$

for the description of the uniaxial flow stress response in terms of the effective von Mises stress σ_e and the accumulated effective plastic strain ε_p; the evolution parameter ε_b in the evolution of b (the initial value of b_o is taken to be 0); the (constant) plastic spin parameter ρ. With only one exponential term in (8), i.e. $K = 1$, a reasonable fit can be obtained to either the small strain or the large strain response; but in order to describe both ranges well, at least two terms will be necessary.

Since cross-sections of the bar can be considered to remain plane and the behaviour during large strain torsion remains axisymmetric, the state of stress and strain inside the bar are uniform in the axial and circumferential directions; but there remains an inherent dependency on the radial coordinate. This problem was solved by means of special purpose one-dimensional (along the radial coordinate) finite elements.

Let us consider now the formulation of a much simpler mechanistic model of this free-end torsion problem to be used as the simplified functions $\tilde{R}_k^\alpha(\boldsymbol{x}, \boldsymbol{a}^\alpha, \tau^\alpha)$ in (6) where $\alpha = 1$ corresponds to the torque response and $\alpha = 2$ to the axial elongation response as functions of the twist angle φ (the loading history parameter). We recall that the approximate models do not need to be the same for the whole range of the material parameters $\boldsymbol{x}$ but will be constructed iteratively. Therefore, the approximate models need only to inherit a functional dependence on the material parameters $\boldsymbol{x}$ similar to the actual (finite element) one, while the suitable set of tuning parameters $\boldsymbol{a}^\alpha$ should allow for a numerically accurate representation of the response functions within a relative small neighbourhood of chosen material parameters.

An approximate model that satisfies the requirements put forward earlier is based on the introduction of a tubular specimen instead of a solid bar. Then the mean radius a_1 and the wall thickness a_2 are considered as the first two of the set of tuning parameters. The introduction of the tube specimen allows to make an assumption that the shear stress is uniform in the radial direction and, introducing more tuning parameters (in total, ten for the torque model and eight for the elongation model), derive the simplified expressions. The expression for the torque model was finally obtained in the explicit analytical form whereas the one for the elongation had to be written in the form of an ordinary differential equation and solved using Runge-Kutta method. The full description of those approximate models is given in Toropov and Van der Giessen (1993).

The application of the identification procedure allowed to find the values of seven material parameters, ε_b, σ_y, $\Delta\sigma_y^{(1)}$, $\varepsilon_o^{(1)}$, $\Delta\sigma_y^{(2)}$, $\varepsilon_o^{(2)}$, ρ, giving the minimum value to the objective function (1). Both response quantities were considered as being equally important for the material parameter identification $(\theta^1 = \theta^2 = 1)$. The Figures 1a and 1b show the comparison of the measured values of torque and axial strain, respectively, and the simulated ones for the obtained set of material parameters.

4.2. THIN METAL SHEET CYCLIC BENDING

It is important to know the material behaviour of thin sheet metals under cyclic loading because the sheets are often subjected to bending-unbending-reversed bending during some metal forming processes. In order to perform an accurate simulation of metal forming, it is necessary to obtain a constitutive model which can describe the cyclic behaviour, especially the deformation-induced anisotropy.

One of the problems in experiments for the determination of cyclic stress-strain curves of thin sheet metals is that the conventional tension-compression test becomes impossible because of buckling of the sheets. Then, instead of tension-compression tests, cyclic bending tests were conducted, and the material parameters for the given constitutive model were identified using the test results presenting the bending moment *vs.* curvature diagrams. A 304 stainless steel thin sheet of 30 mm width and 0.42 mm thickness was subjected to cyclic uniform bending in a certain range of curvature.

In the constitutive model for the infinitesimal deformation (Chaboche and Rousselier, 1983), the cyclic hardening behaviour is described by the isotropic hardening rule, and the Bauschinger effect by the nonlinear kinematic hardening rule, analogous to the hardening rules summarized in Table 1 but with extended evolution rules. A summary of these rules is given in Table 2. The constitutive model incorporates eight material parameters: two elastic constants E and ν ; the initial yield stress σ_y ; parameters Q and b in the isotropic hardening rule; and three parameters a, C and H'_∞ for the kinematic hardening. The calculation of the bending moment for a given curvature was based on the Kirchhoff-Love hypothesis because of the uniform bending condition.

TABLE 2. Summary of hardening rules for sheet bending problem

isotropic hardening rule:	$\sigma_F \equiv H\sigma_y = \sigma_y + R, \quad \dot{R} = b(Q-R)\dot{\varepsilon}_p$
kinematic hardening (KH) rule:	$\dot{\boldsymbol{a}} = \dot{\boldsymbol{a}}_1 + \dot{\boldsymbol{a}}_2$
including nonlinear KH rule	$\dot{\boldsymbol{a}}_1 = C(2a\boldsymbol{D}^p/3 - \boldsymbol{a}_1\dot{\varepsilon}_p)$
and linear KH rule	$\dot{\boldsymbol{a}}_2 = 2H'_\infty \boldsymbol{D}^p/3$

In the identification problem under consideration it was found that the level of numerical noise in the simulation results is such that it does not affect greatly the accuracy of the function values and, therefore, the simple approximation type (7) was used in conjunction with the straightforward SQP algorithm based on the Davidon-Fletcher-Powell update of the approximate inverse of the Hessian matrix. The numerical implementation of the SQP algorithm is based on the routine MINCF by Madsen *et al.* (1990) modified for the use in the batch mode ("black box" approach) when the objective function values are calculated not within the SQP program but by an externally supplied programme for the plane strain analysis of the cyclic sheet bending.

The identification of the above material parameters (excluding Poisson's ratio which was found to be $\nu = 0.3$ from the traditional measurements) was performed using five individual loading and unloading curves (Figure 2) considered as individual response quantities. All five response quantities were considered equally weighted in the formulation of the objective function (1): $\theta^1 = \ldots = \theta^5 = 1$.

Figure 2 illustrates the comparison of the experimental curves of bending moment *vs.* curvature and the corresponding simulated results with the constitutive model incorporating the identified set of material parameters. The simulated curves fit the experimental results reasonably well, though the experimental data include some noise. Figure 3 illustrates the simulated stress-strain curve under uniaxial tension using the identified material parameters; it shows the cyclic hardening characteristics and the Bauschinger effect. The comparison of Figures 2 and 3 illustrates the similarity in cyclic behaviour between the bending moment *vs.* curvature curves and the corresponding stress-strain curves. That is why the proper values of material parameters can be identified from cyclic bending test data inspite of inhomogeneous stress distribution in bending. It should be noted that the identification of material parameters, with which the actual cyclic stress-strain diagrams can be reproduced, is possible when *cyclic bending* experimental data are used but not monotonic bending data only. If the attempt was made to use monotonic bending data only, even the decomposition of the isotropic hardening and the kinematic hardening would be very difficult.

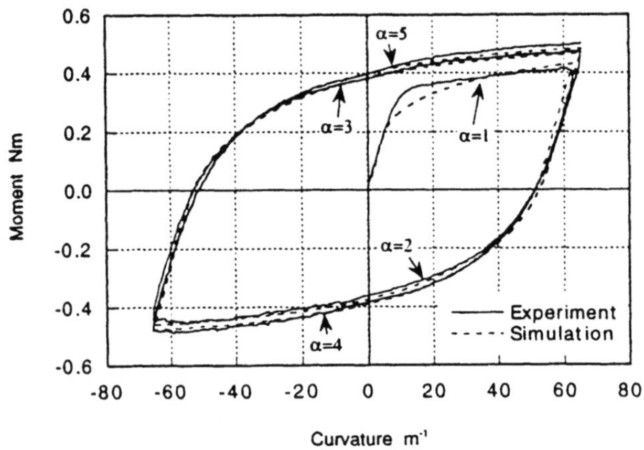

Figure 2. Comparison of experimental and simulated bending moment *vs* curvature curves with the constitutive model incorporating the identified set of material parameters

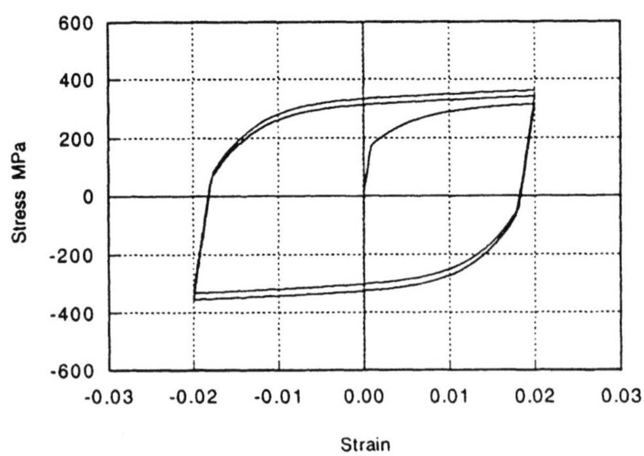

Figure 3. Simulated uniaxial stress-strain curves with the constitutive model incorporating the identified set of material parameters

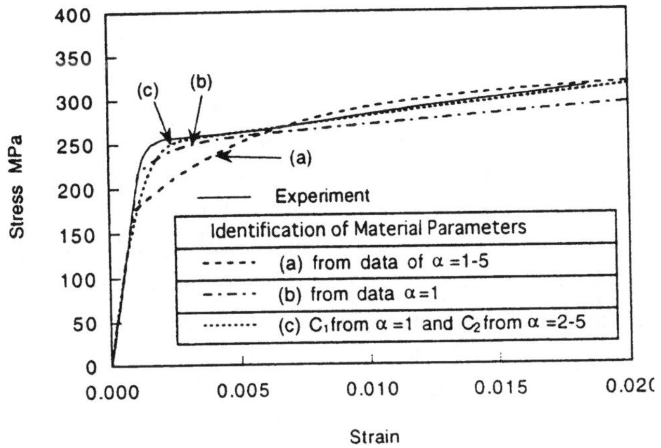

Figure 4. Comparison of the stress-strain curve in uniaxial tension and simulated results when identification was performed (a) using full set of experimental data ($\alpha = 1,..,5$), (b) using the first monotonic bending curve ($\alpha = 1$) and (c) parameter C_1 using the first monotonic bending curve ($\alpha = 1$) and C_2 using subsequent cyclic bending curves ($\alpha = 2,..,5$)

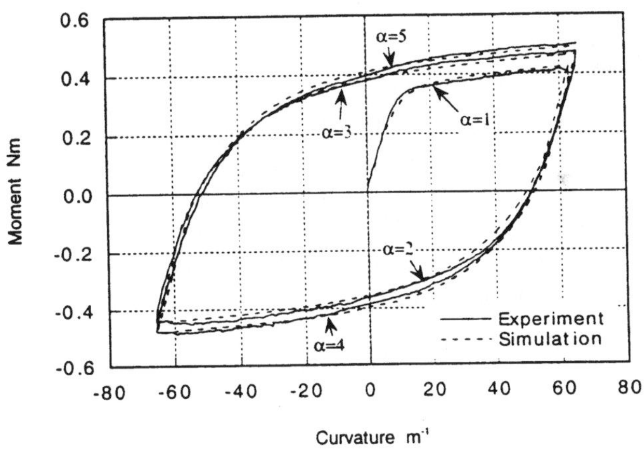

Figure 5. Comparison of the experimental bending moment vs curvature curves and simulated results with the constitutive model incorporating two parameters C_1 identified from the first monotonic bending curve ($\alpha = 1$) and C_2 from subsequent cyclic bending curves ($\alpha = 2,..,5$)

In order to check the accuracy of the identification method, the stress-strain curves, simulated by the constitutive model incorporating the identified material parameters, were compared with the experimental curves obtained from the uniaxial tensile tests (Figure 4, curve (a)). The predicted flow stress values near the initial yield points are found to be lower than the experimental values. This is to be attributed not to the incorrectness of the identification procedure, but to the fact that this type of hardening rule cannot describe the sharp transition of the tangent modulus ($d\sigma/d\varepsilon$) near the initial yield point, when the gradual change of the tangent modulus is simulated in the subsequent reversed (or cyclic) loading (Bauschinger effect). If the material parameters were identified by using the experimental data for the first monotonic bending process only (α=1), the simulated stress-strain curve in uniaxial tension would become closer to the experimental one, as shown in Figure 4 with the line (b). However, the use of thus identified material parameters leads to unacceptably inaccurate prediction of the stress-strain response for the subsequent cyclic loading. The gradual change of the tangent modulus ($d\sigma/d\varepsilon$) in the stress-strain curve corresponding to the gradual change of tangent modulus ($dM/d\kappa$) in the bending moment vs. curvature curve, is expressed by the constitutive model with the small value of initial yield strength σ_y, and large value of a and small value of C in the non-linear kinematic hardening rule. As a simple modification of the constitutive model, let us introduce two different values C_1 and C_2 instead of a material constant C in the non-linear kinematic hardening rule, where C_1 is to be identified from the data for the initial monotonic loading (α=1) and C_2 for the subsequent cyclic loading (α =2, 3, ...), respectively. In Figure 4, the line (c) shows the simulated stress-strain curve. Figure 5 shows the bending moment *vs.* curvature curves simulated with the constitutive model incorporating thus identified parameters C_1 and C_2. In these figures, the simulated results for both the bending and the uniaxial tension agree fairly well with the corresponding experimental results.

5. Concluding Remarks

The presented approach allows for the determination of the phenomenological coefficients in nonlinear constitutive models for engineering materials by minimizing the difference between certain response quantities measured by means of laboratory experiments and obtained by numerical simulation. In principle, it is applicable to any nonlinear constitutive model and is capable of almost unlimited variety of non-trivial experiments, allowing for inhomogeneous material response, the use of structural components as specimens, etc.

6. References

Box, G.E.P. and Draper, N.R. (1987) *Empirical model-building and response surfaces*, Wiley.

Chaboche, J.L. and Rousselier, G. (1983) On the plastic and viscoplastic constitutive equations - Part 1: Rules developed with internal variable concept, *Transactions of ASME, J. of Pressure Vessel Technology*, **105**, 153-158.

Madsen, K., Tingleff, O., Hansen, P.H. and Owczarz, W. (1990) *Robust subroutines for non-linear optimization.* Report NI-90-06, Numerisk Institut, The Technical University of Denmark, Lyngby, Denmark

Marti, K. (1991) Stochastic programming: Numerical solution techniques by semi-stochastic approximation methods, in Slowinski, R. and Teghem, J. (eds.), *Stochastic Versus Fuzzy Approaches to Multiobjective Mathematical Programming under Uncertainty*, Kluwer, pp. 23-43.

Toropov, V.V., Filatov, A.A. and Polynkin A.A. (1993) Multiparameter structural optimization using FEM and multipoint explicit approximations, *Structural Optimization*, **6**, 7-14.

Toropov, V.V. and Van der Giessen, E. (1993) Parameter identification for nonlinear constitutive models: Finite Element simulation - Optimization - Nontrivial experiments,, in Pedersen, P. (ed.), *Optimal Design with Advanced Materials*, Elsevier, pp. 113-130.

Wu, P.D. and Van der Giessen, E. (1991) Analysis of elastic-plastic torsion of circular bars at large strains, *Arch. Appl. Mech.*, **61**, 89-103.

AN INVERSE APPROACH FOR THE IDENTIFICATION OF COMPLEX MATERIAL BEHAVIOURS

O. GHOUATI and J.C. GELIN
Laboratoire de Mécanique Appliquée R. Chaléat
Université de Franche-Comté
24 rue de l'Epitaphe, 25030 Besançon Cédex - France

Abstract : The paper deals with the identification of material parameters for complex behaviours. The strategy proposed relies on the coupling of the finite element method with an optimization technique. The finite element method allows to evaluate the response of a material under various test conditions and to avoid the use of approximation techniques for the interpretation of the experimental results. The proposed method is applied to the determination of an anisotropic behaviour for aluminium alloys.

1. Introduction

Material identification is a major step in behaviour modelling as on it depends the practical use of any established mathematical model for numerical simulations and eventually design optimization of processes involving the use of the considered material. This stage can affect to a great extent the accuracy of the numerical simulations results and it is therefore necessary to pay great attention to the methods used to achieve such identification. Material parameters are identified in a way that they lead to a matching of the calculated response using the model with the measured one, considering the same loading conditions. Therefore, it is necessary to use a numerical method allowing to reproduce faithfully the various thermo-mechanical conditions that can be met in experimental tests. This can be accomplished with the finite element method (FEM) which has reached a certain level of advancement making it a rather accurate tool to simulate a large number of loading conditions and providing a great freedom in the choice of the experimental tests that can be used. Moreover, this technique allows to compare directly the observed response with the calculated one without any conversion or approximation.

Recently, many studies have dealt with the use of the FEM coupled with an optimization technique allowing to adjust material parameters so that the calculated response and the measured one match in a least squares sense [1-3]. These studies are mainly concerned with complex behaviours, seeking better accuracy in the identified material parameters.

H. Sol and C.W.J. Oomens (eds.), Material Identification Using Mixed Numerical Experimental Methods, 93-102.

The technique presented in this paper is suitable for various complex behaviours involving non-linearities and a large number of unknown parameters. Material identification is stated as an inverse problem associated with the direct one consisting of the solution of the mechanical problem with the FEM. Along with the FEM, a modified Levenberg-Marquardt method is used to perform the numerical optimization. This method takes into account constraints on material parameters so that the identified values stay in a range consistent with physical considerations. The Levenberg-Marquardt method, as a gradient based one, requires the evaluation of the calculated response gradient with respect to material parameters. This is accomplished using a sensitivity analysis based on a direct differentiation method (DDM) in order to save computational time and increase the accuracy of sensitivity informations. Special attention is paid to the formulation of the DDM so that it can be consistent with the FEM formulation. Sensitivity data, apart from making the numerical optimization possible, can be helpful for determining confidence intervals on the identified parameters, and for providing a way to choose the experimental test or tests to be performed for a correct identification of all parameters.

An application to the study of deep drawing of aluminium alloys is presented. This study consists of the identification of the anisotropic behaviour of aluminium alloys using one or a combination of experimental tests.

2. Solution of the direct problem

The plastic properties of the sample considered are either isotropic, described by the von Mises yield condition, or anisotropic described by using a rotated frame subjected to follow the anisotropy axes. The hardening can be isotropic with the associated flow rule, but also anisotropic including a linear kinematic hardening with the Prager's rule.

Consider a sample S occupying a domain Ω_0 with boundary Γ_0 in the initial configuration and the domain Ω with boundary Γ in the current configuration. On boundaries Γ_u and Γ_σ ($\Gamma_\sigma \cap \Gamma_u = \emptyset$, $\Gamma_\sigma \cup \Gamma_u = \Gamma$), the displacement $\bar{\mathbf{u}}$ and the traction $\bar{\mathbf{f}}_s$ are prescribed. The sample undergoes a deformation process specified by the displacement field $\mathbf{u}_t(\mathbf{p},t)$ where t is time or a time like parameter for the quasistatic process and $\mathbf{p}$ is the vector of process parameters. In a finite deformation process the initial configuration C_0 is deformed into C_t with $\mathbf{x}_t = \mathbf{x}_0 + \mathbf{u}_t$.

The following weak form associated with the equilibrium equations for the sample S is used in the finite element method:

$$G(\mathbf{u}_t(\mathbf{p}), \eta) = \int_{\Omega_t} \mathbf{T}(\mathbf{u}_t(\mathbf{p})) : \nabla^s_{x_t} \eta \, dV - \int_{\Gamma_\sigma} (\bar{\mathbf{f}}_s \cdot \eta) \, dS = 0 \tag{1}$$

where $\mathbf{T}$ is the Cauchy stress tensor, η is a homogeneous displacement field ($\eta=0$ over Γ_u) and $\nabla^s_{x_t} \eta$ is the symmetrical gradient of η in the current configuration.

Equation (1) is non-linear with respect to $\mathbf{u}_t(\mathbf{p})$. The solution process is performed using a Newton iterative scheme and leads to the following equation:

$$\int_{\Omega_t} \mathbf{C}(\nabla^s_{x_t}\mathbf{u}_t(\mathbf{p})):\nabla_{x_t}\delta\mathbf{u}_t(\mathbf{p}):\nabla^s_{x_t}\eta\, dV + G(\mathbf{u}_t(\mathbf{p}),\eta) = 0 \tag{2}$$

where $\delta\mathbf{u}$ represents the displacement increase. The term $\mathbf{C}(\nabla^s_{x_t}\mathbf{u}_t(\mathbf{p}))$ represents the tangent operator consistent with the algorithm of evaluation of stress and internal variables.

In the displacement based finite element method, the discretized form of the equation (2) is used to calculate an estimated incremental displacement $\delta\mathbf{u}_{n+1}$ between t_n and t_{n+1}, following the implicit iterative equation:

$$[\mathbf{K}_T]^{(i)}_{n+1}\delta\mathbf{u}^{(i)}_{n+1} = [\mathbf{F}_{ext}] - [\mathbf{F}_\sigma(\nabla^s_{n+1}\mathbf{u}_t(\mathbf{p}))]^{(i)}_{n+1} \tag{3}$$

$[\mathbf{F}_{ext}]$ represents the nodal vector of external loads, $[\mathbf{F}_\sigma(\nabla^s_{n+1}\mathbf{u}_t(\mathbf{p}))]$ the nodal vector of internal loads, and $[\mathbf{K}_T]$ represents the tangent stiffness matrix.

In the case of the transient explicit solution procedure, accelerations are taken into account, the virtual work principle can be written in an Eulerian manner as:

$$\int_{\Omega_t} \mathbf{T}(\mathbf{u}_t(\mathbf{p})):\nabla^s_{x_t}\eta\, dV - \int_{\Gamma_\sigma} (\bar{f}_s\cdot\eta)\, dS + \int_{\Omega_t} \rho\gamma\eta\, dV = 0 \tag{4}$$

After discretization, the integral form (5) is transformed as:

$$[\mathbf{M}]_{n+1}\gamma_{n+1} = [\mathbf{F}_{ext}] - [\mathbf{F}_\sigma(\nabla^s_{n+1}\mathbf{u}_t(\mathbf{p}))]_{n+1} \tag{5}$$

where γ_{n+1} is the nodal acceleration vector and $\mathbf{M}_{n+1}$ the mass matrix.

Nodal accelerations vector is obtained by solving equation (6) and velocities, respectively displacements are deduced using a central difference formulae. In the particular case where **M** is a diagonal matrix, the solution process is straightforward.

3. Solution of material identification as an inverse problem

3.1. NONLINEAR OPTIMIZATION

The solution of the parameter estimation problem involves adjusting the parameters in a finite element model until the calculated response, $\mathbf{R}^c_t(\mathbf{u}_t(\mathbf{p}))$ matches the measured one, $\mathbf{R}^m_t$, in the least squares sense [2]. The agreement between the data and the model with a particular set of parameters is measured using the following cost function:

$$S(\mathbf{p}) = \frac{1}{2}\sum_{i=1}^{Nm}\sum_{j=1}^{Nt} [\mathbf{r}_{i,j}(\mathbf{u}_t(\mathbf{p}))]^2 = \frac{1}{2}\mathbf{r}(\mathbf{u}_t(\mathbf{p}))^T\mathbf{r}(\mathbf{u}_t(\mathbf{p})) \tag{6}$$

where Nm is the number of measurements performed at Nt values of the discrete time. The residual vector $\mathbf{r}(\mathbf{u}_t(\mathbf{p}))$ is defined by:

$$\mathbf{r}(\mathbf{u}_t(\mathbf{p})) = \mathbf{R}_t^c(\mathbf{u}_t(\mathbf{p})) - \mathbf{R}_t^m \tag{7}$$

The strategy developed is to take profit from the incremental aspect of the finite element method in the nonlinear range. This provides the necessary number Nm for the values of the unknown material parameters to be identified. For each value t_j, $j=1,\ldots,Nt$, of the discrete time (loading), a material response, $\mathbf{R}_t^c(\mathbf{u}_t(\mathbf{p}))$, is evaluated using the model and a measured response, $\mathbf{R}_t^m$, is deduced from the loading curve obtained from the experimental test.

The cost function S is modified to take account for physical constraints on material parameters by the definition of the function S^* as:

$$S^*(\mathbf{p}) = S(\mathbf{p}) + \sum_{j=1}^{q} \frac{\omega_j}{\mathbf{C}_j(\mathbf{p})} \tag{8}$$

where ω_j are non-negative weights, and $C_j(\mathbf{p})$ are the q constraint functions ($\mathbf{C}_j(\mathbf{p}) \geq 0 \quad j = 1, q$). In the case of independent material parameters identification, the constraint functions C_j are linear functions in p, used to limitate the variation domain of the parameters. The computation of the derivatives of these functions is therefore possible.

A Levenberg-Marquardt method is used to minimize the modified cost function following the equation [2,4]:

$$\left[(\mathbf{J}^{(k)})^T\mathbf{J}^{(k)} + \lambda^{(k)}\mathbf{I} + \mathbf{H}^{(k)}\right]d\mathbf{p}^{(k)} = -(\mathbf{J}^{(k)})^T\mathbf{r}^{(k)} + \mathbf{g}^{(k)} \tag{9}$$

where λ is the Levenberg-Marquardt parameter (a non-negative scalar), **J** is the jacobian matrix of S, and **g** and **H** contain the first and second derivatives, respectively, of the weighted penalty functions with respect to the parameters.

The components of the jacobian matrix **J** are given by the following expression:

$$J_{i,j}^{\alpha} = \frac{\partial r_{i,j}}{\partial p_\alpha} = \frac{\partial R_{i,j}}{\partial p_\alpha} \quad ; \quad \begin{matrix} i = 1, Nm \\ j = 1, Nt \end{matrix} \quad ; \quad \alpha = 1, n \tag{10}$$

3.2. SENSITIVITY ANALYSIS

The evaluation of the jacobian matrix requires a sensitivity analysis, which can be performed by different means. The most straightforward and the easiest one to implement is the finite difference method, but it can also be computationally expensive and deficient in terms of accuracy and reliability. In this paper, we will use the direct differentiation method which provides very accurate sensitivity informations if the formulation of the sensitivity analysis is consistent with the formulation of the finite element method [5-1]. Terms to determine are given by equation (10).

By using the fact that the displacements are the result of an incremental step-by-step calculation, these terms can be rewritten as follows:

$$J^{\alpha}_{i,j} = \sum_{n=0}^{k-1} \left(\frac{d\,(\Delta u_{i,j}^{(n+1)})}{dp_{\alpha}} \right) \tag{11}$$

Next for simplicity of notations, we will drop the indices i, j.

The increment of displacement Δu_{n+1} is obtained when the equilibrium is satisfied, which from equation (3) can be written as:

$$\mathfrak{R}_{n+1} = [\mathbf{F}_{ext}] - [\mathbf{F}_{\sigma}(\nabla^{s}_{n+1}\mathbf{u}_t(\mathbf{p}))]^{(i)}_{n+1} = 0 \tag{12}$$

in the case of dead external loading.

The total differentiation leads to:

$$\frac{d\mathfrak{R}_{n+1}}{d\mathbf{p}} = \frac{\partial}{\partial \mathbf{p}}[\,[\mathbf{F}_{\sigma}(\nabla^{s}_{n+1}\mathbf{u}_t(\mathbf{p}))]_{n+1}] + [\mathbf{K}_T]_{n+1}\left(\frac{d\,(\Delta \mathbf{u}_{n+1})}{d\mathbf{p}}\right) = 0 \tag{13}$$

and solving this equation for the unknowns yields:

$$\left(\frac{d\,(\Delta \mathbf{u}_{n+1})}{d\mathbf{p}}\right) = -[\,[\mathbf{K}_T]_{n+1}]^{-1}\left[\frac{\partial}{\partial \mathbf{p}}[\,[\mathbf{F}_{\sigma}(\nabla^{s}_{n+1}\mathbf{u}_t(\mathbf{p}))]_{n+1}]\right] \tag{14}$$

The stiffness matrix $[\mathbf{K}_T]_{n+1}$ is available from the solution of the direct problem. It remains to determine the derivative of the internal load vector as:

$$\frac{\partial}{\partial \mathbf{p}}[\,[\mathbf{F}_{\sigma}(\nabla^{s}_{n+1}\mathbf{u}_t(\mathbf{p}))]_{n+1}] = \mathop{\mathbf{A}}_{\mathbf{e}} \int_{\Omega^{e}_{n+1}} (\beta^{e\,(i)}_{n+1})^{T} \frac{d}{d\mathbf{p}}[\mathbf{T}_{n+1}(\mathbf{u}_t(\mathbf{p}))]\, dV^{e} \tag{15}$$

The sensitivity of the Cauchy stress tensor $\frac{\partial}{\partial \mathbf{p}}[\mathbf{T}_{n+1}]$ appearing in equation (15) is evaluated in a manner that is consistent with the algorithm of stress and state variables calculation, following a scheme similar to the one used to evaluate the Cauchy stress tensor [6,7].

4. Application to the determination of aluminium alloys material parameters

The flow law chosen to represent the behaviour of aluminium alloys studied here is a Voce law of the following form:

$$\sigma_f = \sigma_Y + Q\,(1 - \exp(-b\varepsilon^p)) \tag{16}$$

where σ_f and ε^p represent respectively the flow stress and the equivalent plastic strain.

To represent anisotropy, the quadratic model of Hill is used. The equivalent stress is then expressed by the following relation:

$$\bar{\sigma} = \mathbf{H}:\mathbf{T}:\mathbf{T} \tag{17}$$

where **T** is the Cauchy stress tensor and **H** is the Hill operator, a fourth order tensor, with the following matricial expression:

$$\mathbf{H} = \begin{bmatrix} G+H & -H & 0 \\ -H & F+H & 0 \\ 0 & 0 & 2N \end{bmatrix} \tag{18}$$

The Hill coefficient F, G, H and N can be expressed in terms of the Lankford coefficients as following:

$$H = \frac{r_0}{1+r_0} \;;\; F = \frac{r_0}{r_{90}(1+r_0)} \;;\; N = \frac{(r_0+r_{90})(2r_{45}+1)}{2r_{90}(1+r_0)} \quad \text{and } G+H = 1 \tag{19}$$

For a sample loaded in a direction with an angle θ with the rolling direction, the components of the Hill matrix can be expressed as a function of F, G, H, N and the cosines of the angle θ.

Material parameters vector to identify is therefore the following:

$\mathbf{p}=[p_1,p_2,p_3,p_4,p_5,p_6]=[\sigma_Y,Q,b,r_0,r_{45},r_{90}]$.

The simple bounds on the parameters, reported in Table 1, were introduced in the process of identification.

TABLE 1. Parameter constraints

$0 < \sigma_Y < 500$
$0 < Q < 500$
$0 < b < 50$
$0 < r_0 < 10$
$0 < r_{45} < 10$
$0 < r_{90} < 10$

Material parameters identification was performed using tensile tests.

For the numerical simulations an exact geometrical representation of the tensile specimen was used. The geometry and the mesh used are represented in figure 1, with the deformed mesh obtained after a direct analysis.

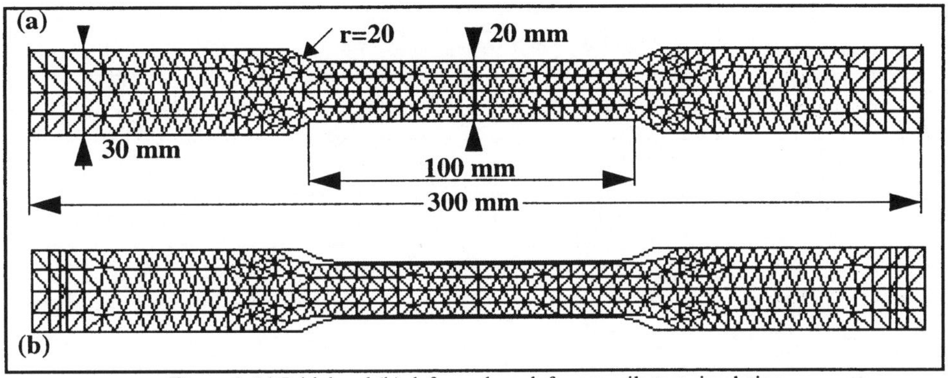

Figure 1. (a) Initial and (b) deformed mesh for a tensile test simulation

Next we present the results obtained for the identification first assuming an isotropic material and then an anisotropic material.

4.1. ISOTROPIC CASE

In this part, we assume that the material is isotropic and the identification is performed to determine the three material parameters: σ_Y, Q, b.

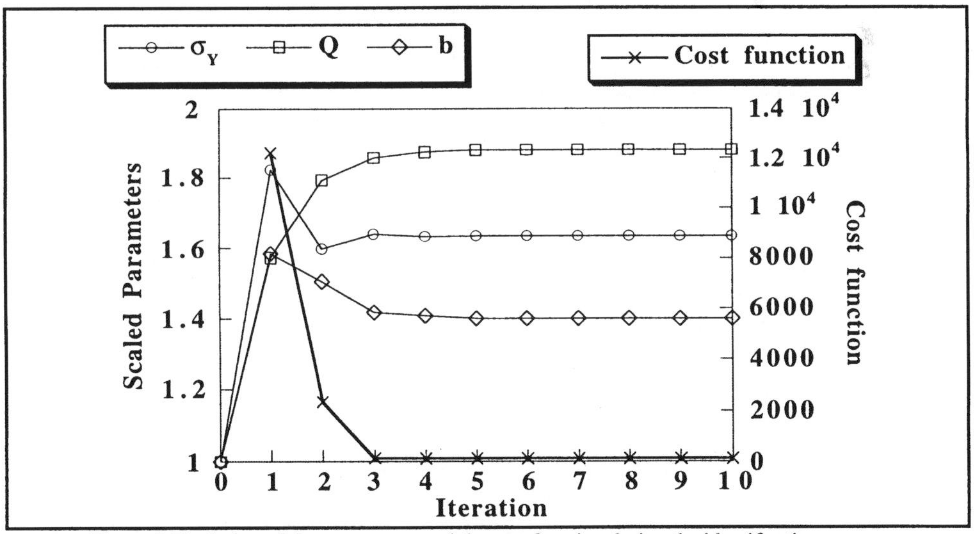

Figure 2. Evolution of the parameters and the cost function during the identification process

Figure 2 represents the evolution of material parameters and the cost function during the identification process. It can be noticed that the iterative process converges rapidly to the solution (4 iterations). Based on the results of the sensitivity analysis, uncertainties on the calculated parameters can be evaluated. These uncertainties are provided in table 2 with the results of the identification and the initial guesses for the parameters.

TABLE 2. Initial and identified values

Parameter	Initial value	Identified value	Uncertainty
σ_Y	100	163.4	0.005
Q	100	188.1	0.006
b	10	14	0.016

4.2. ANISOTROPIC CASE

In this part, the anisotropy of the material is taken into account and the identification is performed to determine the six material parameters. To be able to perform this identification, we used three tensile tests, for $\theta=0°$, $\theta=45°$ and $\theta=90°$.

Figure 3 shows the evolution of material parameters and the cost function during the identification process whereas in table 3, the identified values of the parameters are reported. It can be noticed that the values of the parameters σ_Y, Q and b are not much changed when considering an anisotropic material.

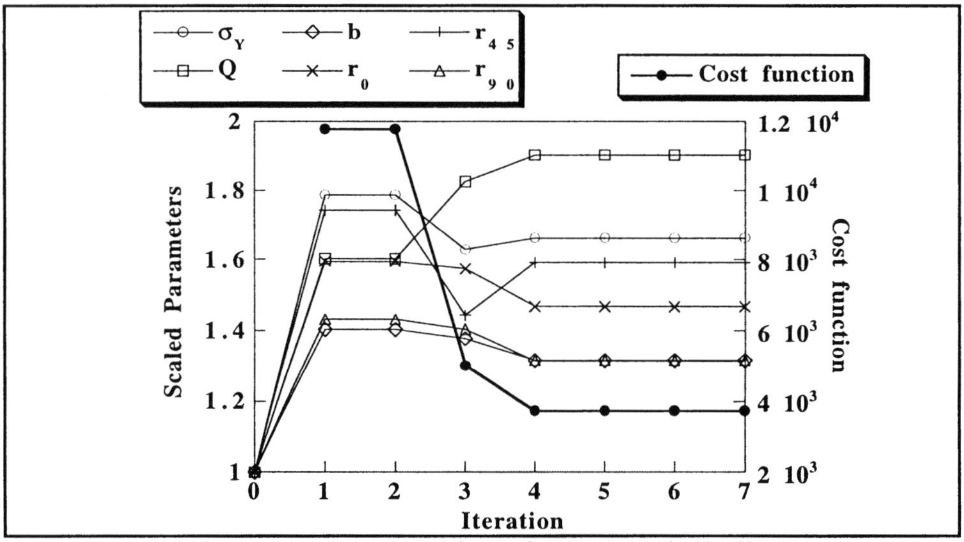

Figure 3. Evolution of the parameters and the cost function during the identification process

Uncertainties on parameters r_0, r_{45} and r_{90} are relatively high, showing that these parameters are not well identified. This can be due to two main reasons. First one is the fact that the identification is based on the hardening curve which might not be very sensitive to these parameters. One should include geometrical aspects in all directions which will exhibit more sensitivity. Second reason is the fact that the identification uses only a set of experimental tests and no information on data errors is taken into account.

TABLE 3. Initial and identified values

Parameter	Initial value	Identified value	Uncertainty
σ_Y	100	166.2	0.11
Q	100	190.2	0.13
b	10	13.2	0.01
r_0	1.0	1.469	0.93
r_{45}	1.0	1.592	0.95
r_{90}	1.0	1.315	0.78

4.3. ON SENSITIVITY ANALYSIS

As stated earlier a direct differentiation method was used to perform sensitivity analysis. It is interesting to compare the results obtained using this method and those obtained using a finite difference method.

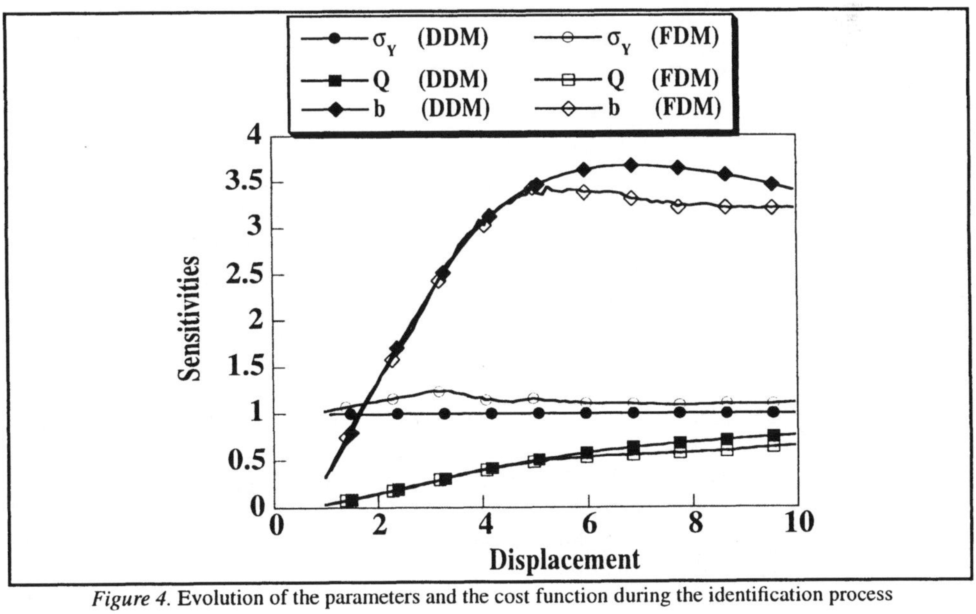

Figure 4. Evolution of the parameters and the cost function during the identification process

O. GHOUATI

TABLE 4. Initial and identified values

Parameter	Initial value	DDM	FDM
σ_Y	100	163.4	172.7
Q	100	188.1	206.2
b	10	14	9.5

Figure 4 shows a comparison of the sensitivities of the response with respect to the three parameters σ_Y, Q and b, at the beginning of the identification. The sensitivities provided by the two methods are different, and this affects the results of the identification as it can be seen in Table 4. The set of parameters reached using the FDM does not lead to a minimum of the cost function showing that it is not the correct solution.

5. Conslusions

In this paper, a technique for the identification of material parameters for complex behaviour was presented. This technique is based on the combination of a finite element method and an optimization method. The finite element method is used to evaluate the response of the material subject to various loadings for different experimental conditions. This identification technique was successfully applied to the determination of the material parameters for an anisotropic behaviour, using tensile tests in different directions. Current works concern the correct identification of anisotropic coefficients and the identification of material parameters directly using a deep drawing test.

6. References

1. Gavrus, A., Massoni, E. and Chenot, J.L.: Computer aided rheology for non linear large strain thermo-viscoplastic behaviour formulated as an inverse problem, *Inverse Problems in Engineering Mechanics*, Bui, Tanaka *et al.* (eds), Balkema (1994), pp. 123-130.

2. Gelin, J.C. and Ghouati, O.: An Inverse Solution Procedure For Material Parameters Identification In Large Plastic Deformations, *Comm. in Num. Meth. in Eng.* **12** (1996), 161-173.

3. Gelin, J.C. and Ghouati, O.: An inverse method for material parameters estimation in the inelastic range, *Computational Mechanics* **16** (1995), 143-150.

4. Schnur, D.S. and Zabaras, N.: An inverse method for determining elastic material properties and material interface, *Int. J. Numer. Methods. Eng.* **33** (1992), 2039-2057.

5. Vidal, C.A., Lee, H.S. and Haber, R.B.: The consistent tangent operator for design sensitivity analysis of history-dependent response, *Comp. Syst. in Eng.* **2** (1991), 509-523.

6. Leu, L.J. and Mukherjee, S.: Implicit objective integration for sensitivity analysis in non-linear solid mechanics, *Int. J. Numer. Methods. Eng.* **37** (1994), 3843-3868.

7. Ghouati, O. and Gelin, J.C.: Inverse Identification Methods For Material Parameters Estimation in Metal Forming, *Simulation of Materials Processing: Theory, Methods and Applications*, NUMIFORM'95, S.F. Shen and P. Dawson (eds), Balkema (1995), pp. 569-574.

DISCUSSION ON SOME ESTIMATORS IN MIXED NUMERICAL/EXPERIMENTAL TECHNIQUES

H. HUA* H. SOL** W.P. DE WILDE**
*Shanghai Jiao Tong University
1954 Hua-Shan Rd., 200030 Shanghai
**University of Brussels
Pleinlaan 2, 1050 Brussels

ABSTRACT The mixed numerical/experimental techniques (MNET), in which the numerical results are correlated with experimental data in an iterative fashion, has been developing for more than two decades. An overview of the various estimation algorithms in the MNET is given. Emphasis is placed on the different philosophies which may lead to a deterministic optimisation procedure, or a statistical estimator, or a combination of the two. Formulations, limitations and internal relations of the algorithms are also presented. All the discussions are oriented to the researchers and engineers who are interested in the design optimisation, parameter identification and finite element model updating.

1. Introduction

The concepts and applications of the mixed numerical/experimental techniques for the identification of dynamic characteristics of a given structure, and the refinement of a finite element model dated from two decades ago [1,2,3]. Two fundamental models are involved: experimental modal model, and numerical model (usually a finite element model). Using an appropriate estimation algorithm which is in general in an iterative fashion, the MNET updates a set of parameters of the numerical model by correlating the numerical results with the experimental results. By deterministic optimisation, the starting point is to construct an objective function

$$J(x) = \{\Delta y\}^T[W_y]\{\Delta y\} + \{\Delta x\}^T[W_x]\{\Delta x\} \tag{1}$$
$$\{\Delta x\} = \{x_a\} - \{x_o\}$$
$$\{\Delta y\} = \{y_a\} - \{y_e\}$$

in which $\{x_a\}$is nx1 parameter vector to be updated, $\{x_o\}$are nx1 initial parameters, $\{y_a\}$and , $\{y_e\}$are mx1 analytical value vector and experimental value vector respectively. $[W_y]$ and $[W_x]$ are positive definite, usually also symmetric weighting matrices. By statistical estimation, the conditional probability density function (equation (2)), or joint probability density function (equation (3)) is maximised. Stated mathematically, that is [4]:

$$\max_x p(\{Y\}|\{x\}) = p(\{Y\}|\{\hat{x}\}) \tag{2}$$

$$\max_x p(\{Y\},\{x\}) = p(\{Y\}, \{\hat{x}\}) \tag{3}$$

in which $\{Y\}$ is the observation of $\{y_e\}$ and $\{\hat{x}\}$ is the estimation of $\{x\}$. This leads to conditional or unconditional maximum likelihood estimation. The other statistical estimation is based on the minimisation of a quadratic cost function [4]:

H. Sol and C.W.J. Oomens (eds.), Material Identification Using Mixed Numerical Experimental Methods, 103-108.

$$\min_{x_{est}} E\left[\{x - x_{est}\}^T [W] \{x - x_{est}\}\right]$$
$$= E\left[\{x - \hat{x}\}^T [W] \{x - \hat{x}\}\right] \tag{4}$$

in which $\{x_{est}\}$ is the estimation of $\{x\}$ given $\{Y\}$, and $[W]$ is a positive definite symmetric weighting matrix. A very interesting phenomenon is that, starting from either an objective function, or probability density function, or some cost function, under certain circumstances, they may arrive at the same destination. However, under other conditions, they may result in completely different results. It is evident that the algorithm resulted from any one of the above equations from (1) to (4) is optimal, since the solution is obtained either by minimisation or by maximisation. Therefore, the criterion of optimality should be provided for the users to avoid fuzziness in the selection of estimation algorithms and, frequently, selection one from too many optimal solutions. Probably, not the mathematical explanation, but a concise philosophy explanation is the better way to clear the vagueness. So, this paper first presents the formulations of the most frequently used estimators, then explains the optimality criteria, convergence criteria and the philosophy on which each estimator based; Finally, practical applications of these estimators demonstrate their robustness.

2. Theoretical background

Three fundamental algorithms are presented: deterministic optimisation, extended maximum likelihood estimation , and Kalman filter.

(1) deterministic optimisation using modified method of searching in feasible direction. Although there are many deterministic optimisation method, there is no difference in philosophy. So, only the method searching in usable and feasible direction is briefly reviewed. The mathematical model of design optimisation can be stated as

$$\min J(X),\ x_i^l < x < x_i^u,\ \ i=1,2,\ldots n \tag{5}$$
$$\text{subject to } g_j(x) \le 0,\ j = 1,2,\cdots k$$

where J(x) is the equation(1), $g_j(x)$ are constrains, for example, total mass, x^l and x^u are lower and upper bounds of the design variables. Given an initial vector $\{x_o\}$, it updates the design variables by searching the optimal scalar parameter α^* in usable-feasible direction S^k [9]

$$x^k = x^{k-1} + \alpha^* S^k \tag{6}$$

The searching direction might be the steepest decent direction, or the Ftecher-Reeves conjugate direction, or other various strategies. If α^* corresponds to encountering the bound of some variable x_k, Kuhn-Tucker condition should be satisfied. Then fix this value at its bound and repeat the search from this point with the remaining variables, find a new searching direction, new α^*,.... Convergence criterion can be either $|x^k - x^{k-1}| < \varepsilon$ or $|y^k - y^{k-1}| < \varepsilon$ or combination of the two.

(2) Extended maximum likelihood estimation, a Bayes type estimator [4] The stochastic form of the model is given by

$$\{y\} - \{y_o\} = [S](\{x\} - \{x_o\}) \tag{7}$$
$$\{y_e\} = \{y\} + \{\varepsilon\} \tag{8}$$

setting

$$\{\Delta y\} = \{y_e\} - \{y\},\ \ \{\Delta x\} = \{x\} - \{x_o\} \tag{9}$$

the statistical properties(mean and variances) of $\{\Lambda x\}$ and $\{\Lambda y\}$are assumed to be normal

$$E\{\Delta x\}=E\{\Delta y\}=0, \quad E(\{\varepsilon\}\{\Delta x\}^T)=0 \tag{10}$$

$$E(\{\Delta x\}\{\Delta x\}^T)=V_x, \quad E(\{\varepsilon\}\{\varepsilon\}^T)=V_y \tag{11}$$

In the mathematical model, $\{\varepsilon\}$ is the measurement noise which is assumed to be white gaussian and uncorrelated with $\{\Lambda x\}$. Iteration starts from initial parameter and variance, and ends when exact $\{y\}$ is obtained (in this case, $\{\Lambda y\}$---->$\{\varepsilon\}$), or within a certain accuracy. The joint probability density function of $\{\Lambda y\}$ and $\{\Lambda x\}$ can be written as (vector signs { } [ ] are omitted)

$$p(\Delta x, \Delta y)=\frac{1}{(2\pi)^{\frac{n+m}{2}}\sqrt{V_x V_y}}\exp-\frac{1}{2}(\Delta x^T W_x \Delta x+\Delta y^T W_y \Delta y)$$

$$W_x=V_x^{-1}, \quad W_y=V_y^{-1} \tag{12}$$

$[W_x]$ is given by more or less subjective guess, $[W_y]$ is independent of $\{x\}$, therefore, maximise the joint probability density function is equivalent to minimise a cost function equation (1). The minimisation of (1) or maximisation of (12) leads to the updating algorithm

$$\begin{aligned}\{x\}&=\{x_o\}+[K]\{\Delta y\}\\ [K]&=[V_x][S]^T([S][V_x][S]^T+[V_y])^{-1}\\ [V_x^*]&=([I]-[K][S])[V_x]\end{aligned} \tag{13a}$$

For non-linear model, $[S]=\partial y/\partial x$ is the sensitivity matrix. $[V_x^*]$ is the estimation error covariance matrix. It can be found that with the normal distribution assumption , the estimation is minimum variance estimation.

An equivalent expression is

$$\begin{aligned}\{x\}&=\{x_o\}+[K]\{\Delta y\}\\ [K]&=([S]^T[W_y][S]+[W_x])^{-1}[S]^T[W_y]\{\Delta y\}\\ [V_x^*]&=([W_x]+[S]^T[W_y][S])^{-1}\end{aligned} \tag{13b}$$

(3) Extended Kalman filter [6, 8]

Extended Kalman filter has the similar mathematical expression as maximum likelihood estimator (equation (13)). Extended Kalman filter is the extension of linear Kalman filter to the non-linear system. The linear Kalman filter can be derived by minimisation of equation (4). The statistical properties are not necessary to be normal. But, applied to parameter estimation of a linear system, it must satisfy the observability condition [7]:

$$[S]^T[V_y]^{-1}[S]>0 \tag{14}$$

The estimation error covariance matrix at (n-1)-th iteration is used at n-th iteration. If convergence can be reached, the necessary and sufficient condition for a linear system is obvious: the estimation error should monotonically reduces. That is

$$\left\|V_x^*(k)/V_x^*(k-1)\right\|<1 \tag{15}$$

3. Discussion on the philosophy

Deterministic optimisation By construction, the objective function is convex. So, using design optimisation, the minimum point can always be found. The initial estimation will influence the searching rate. The weighting coefficients $[W_y]$ are determined according to the relative importance of the state variables or the norm of their sensitivities. $[W_x]$ is determined in the same way. In the point of view of design optimisation, no *a priori* knowledge about initial values of the given system has to be taken into account when searching in usable-feasible direction.

Maximum likelihood estimation This type estimation is to choose that value of $\{x\}$ which maximise the joint probability density function evaluated at given observation $\{Y\}$. That is its optimality criterion. It is a Bayes type estimator, it assumes that the entire information available to the estimator is contained in the *a posteriori* probability density function of $\{x\}$ conditioned on observation $\{Y\}$, since only the knowledge of this probability density function is required to estimate the values for cost function. The weighting coefficients $[W_x]$ and $[W_y]$ are determined by their statistical properties.

Kalman filter In the statistical point of view, it has the same philosophy as maximum likelihood estimation, but it minimises the error covariance matrix. It is a minimum variance estimator. The main difference from the maximum likelihood estimator is that the estimation error covariance matrix at (n-1)-th iteration is used at n-th iteration. By control theory, it states, if $[W_x]$ the confidence of the previous estimation is higher, [K] will be smaller, less modification will be made with the parameters, otherwise, more modification will be made with the parameters; and if the measurement error is smaller, $[W_y]$ will be larger, gain matrix [K] will be large too, more modification will be made with the previous estimated parameters. It works like an adaptive estimator. Based on more or less subjective guess of the initial statistical properties, using the adaptive characteristics to modify the previous judgement, is the philosophy of the Kalman filter.

4. Uniqueness and robustness of the solution

Uniqueness of the solution Uniqueness only exists for a linear observable system. The problem may be determined, undetermined, overdetermined depending on the effective parameter dimension and the observation dimension. For undetermined system, or unobservable system in which the rank of the sensitivity matrix is smaller than the number of the parameters to be estimated, Kalman filter is no longer valid (equation (15) can not be satisfied), deterministic optimisation method may get some acceptable solution, and maximum likelihood estimator can certainly obtain some optimal solution in the least square sense, since the pseudo-solution is the special case of Basin estimator, if setting the weightings $[W_y]$ to infinite and $[W_x]$ to unit, we have

$$\{\Delta x\} = [S]^T([S][S]^T)^{-1}\{\Delta y\} \tag{16}$$

or setting $[W_y]$ to unit and $[W_x]$ to zero

$$\{\Delta x\} = ([S]^T[S])^{-1}[S]^T\{\Delta y\} \tag{17}$$

Equation (16) and (17) are the pseudo-solusions.

Robustness of the estimators Either $[W_y]$ to infinite or $[W_x]$ to zero, by the philosophy of Kalman filter, implies a permission of large modification of parameters. So, the Bayesian estimators are more robust than the pseudo-solution. Because of the convexity of the objective function, the optimisation method is, by the experience of the authors, the most robust one, but less accurate. On the other side, with maximum likelihood estimator, the matrix in parenthesis of equation (13) can always be inversed because of the positiveness of the weighting matrices $[W_y]$ and $[W_x]$. Physically, the measurement data is more or less noise corrupted and estimation can never be perfect. This property is unintentionally utilised by Ojavo [11]. Kalman filter is relatively less robust, but more accurate, because of the strict requirement of convergence (equation (15)).

5 Case study

Example 1 An unobservable system updating. The structure (Fig. 1) under study is an oilpan which was constructed using 900 shell elements (NASTRAN CTRIA3 and CQUAD4). Before model updating, the whole model is divided into 29 element groups. Each group of elements are assigned to one physical property entry, in order to get a reasonable amount of design variables. To construct element group is equivalent to determine parameters to be updated, and these element groups should show sufficient controllability of the modes within the frequency band of interest.
The total mass requirement is applied as a constraint in design optimisation, and as one of observations in maximum likelihood estimator. The following scaled objective function was used in the calculation for design optimisation:

$$\min \quad \sum_{i=1}^{m}(\Delta y_i / y_i - 1)^2 W_{yi} + \sum_{j=1}^{n}(\Delta x_j / x_j - 1)^2 W_{xj}$$

$$\text{subject to} \quad m^l < \text{total mass} < m^u$$

And by maximum likelihood estimation, the following cost function is minimised:

$$\min \quad \sum_{i=1}^{m}(\Delta y_i / y_i - 1)^2 W_{yi} + (\Delta m / m - 1) W_m + \sum_{j=1}^{n}(\Delta x_j / x_j - 1)^2 W_{xj}$$

in which m is total mass. $W_{yi}=(0.01)^2$, $W_m=(0.005)^2$, $W_{xi}=(0.025)^2$, with the same lower and upper parameter limits.
Table 1 summarises the resonance frequencies of the model, before and after model updating, using deterministic optimisation and maximum likelihood estimation. First six modes are used in calculation. The parameter change with respect to initial values are show in Fig. 3.

Example 2 Identification of composite material properties, an overdetermined system identification problem. This FE model consists of 192 four-node shell elements and 216 nodes (Fig. 2). The first ten modes were used to estimate four parameters: Young´s modulus E_x, E_y, shear modulus G_{xy} and Poisson´s ratio v_{xy} (thickness and mass density were precisely measured). The upper one of Fig. 4 shows the parameter modification by deterministic optimization and Kalman filter, and the lower one variance evolution by Kalman filter. In the evaluation, a higher confidence was assigned to Poisson´s ratio, and very small modification was obtained for the Poisson´s ratio.

Table 1 Resonance frequency comparison

Test results		before updating		updating using maximum likelihood estimation		updating using deterministic optimisation	
mode No.	Frequency Hz	Frequency Hz	difference	Frequency Hz	difference	Frequency Hz	difference
1	181.9	125.4	31.1	177.0	2.68	175.3	3.6
2	334.6	311.7	6.8	343.9	-2.8	342.3	-2.3
3	513.4	434.5	15.36	521.4	-1.55	501.7	2.28
4	724.6	689.5	4.84	733.6	-1.24	742.7	-2.50
5	837.5	782.2	6.60	814.0	--2.83	824.2	1.58
6	940.4	862.3	8.30	942.6	0.24	949.5	-0.97

6 Conclusion

From example 1, it can be found that, although the frequency differences are reduced for both of the deterministic and probabilistic methods, parameter modifications are very much different. The initial parameter selection dependence of the design optimisation is, of course, one of the important reasons. The other reason is: the problem itself is an undetermined problem, there exist more unknowns than equations. From example 2, it can be found that all of the estimators worked very well. However, maximum likelihood estimation and Kalman filter can provide with not only means but also estimation error variances. For the accurate estimation requirement, Kalman filter and maximum likelihood estimation are superior to deterministic optimisation; It also show that deterministic optimisation in combination with statistical weightings will also yield good results.

Fig. 1 Finite element model of an oilpan

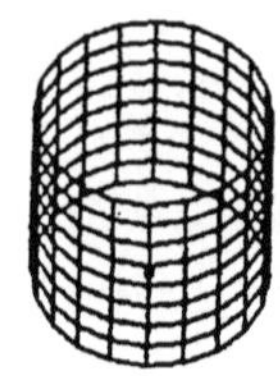

Fig. 2 A composite shell

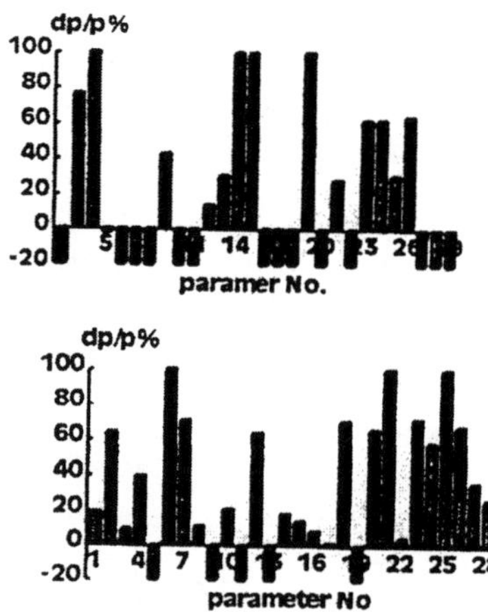

Fig. 3 Parameter modification
Upper one: maximum likelihood estimation
Lower one: NASTRAN optimisation results

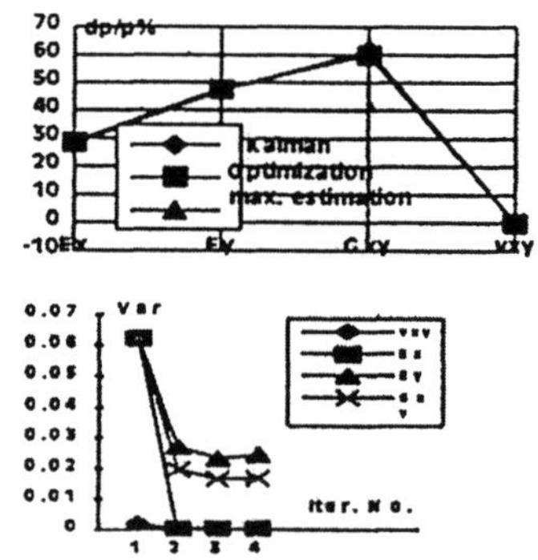

Fig. 4
Upper one: Parameter modification
Lower one: variance evolution by Kalman filter

References

1. Collins, J. et al. (1974) Statistical identification of structures, AIAA 12(2), 185-190
2. Hasselman, T. (1983) A perspective on dynamic verification, Winter annual meeting of ASME, Boston, MA, 119-133
3. Mettershead, J.E., Friswell, M.I.(1993), J. Sound and vibration 167(2), 347-375
4. Nahi, N.E. (1976), Estimation theory and application, E.K. Publishing Co., Huntington, New York, 78-79, 83-84
5. Catin, D.E. (1989), Estimation, control, and discrete Kalman filter, Springer-Verlag
6. Anderson, B.D.O., Moore, J.B. (1979), Optimal filtering, Pretice-Hall, Englewood Clif, New Jersey
7. HUA, H. (1993), Identification of plate rigidities of anisotropic rectangular plates, sandwich panels and circular orthotropic discs using vibration data, PhD dissertation, University of Brussels, 135-146
8. Blakely, K.(1992), Using MSC/NASTRAN to match frequency response test data, Proc. of 11-th IMAC, 42-45
9. Carret, N. Vanderplaats, Numerical optimisation techniques for engineering design, McGraw-Hill Book Co., 1984
10. Colub, G.H., Van Loan, C.F. (1983), Matrix computation, North Oxford Academic
11. Ojavo, I.U.(1990), Improved solution for system identification equation by epsilon-decomposition, 31-th AIAA SDM conference

STOCHASTIC PARAMETER IDENTIFICATION IN THE CASE OF DATA UNCERTAINTIES

I.N. TRENDAFILOVA
Institute of Mechanics, Bulgarian Academy of Sciences,
bl.4, Acad. G. Bontchev str., 1113 Sofia, Bulgaria.

The inverse problem for parameter identification from measurements of the excitation and the response of a system is considered, in the presence of uncertainties in the response data. A method is offered to estimate the statistical characteristics of the model (system) parameters, given the probability distribution of the measured response. The main idea is to offer a general enough method for inverse identification of parameters, that could be applied for a lot of inverse parameter identification problems of interest, provided one can solve the deterministic direct problem. The general formulation of the method is developed for a rather wide class of systems. A system of random functions is introduced to approximate the identification (inverse) problem solution. The inverse problem is solved minimizing an objective function representing the mean value of the difference between the measured and the model responses. The expected values, variances and covariances of the model parameters are evaluated. The developed procedure is applied to the problem 1) for parameter identification of material models from one dimensional dynamic experimental data; 2) for crack identification form electric potential measurements.

Keywords: system identification, stochastic inverse problems, parameter identification, discrete random functions expansion.

1.Introduction.

Uncertainties exist and they must be taken into account in many engineering problems. A typical source of uncertainty lies in the determination of the excitation of the structure. A lot of external loads in practice turn out to be stochastic in nature and they are considered as such. Another source of uncertainty arises from the structure modeling. In the analysis problem all these finally produce model response uncertainties. Measured response uncertainties are also common and they could be due to the experimental measurements, uncertainties in the excitation, boundary conditions etc. When considering the inverse problem, which requires the modeling of the structure given its measured response, the response uncertainties naturally result in model uncertainties. Thus the above two problems are closely related and are an object of common interest. The interrelationship between the statistical characteristics of the model parameters and the statistical characteristics of the response is also a point of general attention.

A number of papers have been published concerning the application of stochastic

H. Sol and C.W.J. Oomens (eds.), Material Identification Using Mixed Numerical Experimental Methods, 109-121.

methods and random field theory in problems for systems with uncertain parameters [1,2,3]. Simulation and perturbation methods were used to study the effect of parameter uncertainties on the response of structural and geotechnical systems [9,10]. Neuman expansion method was applied by Yamazaki, Shinozuka et al. [3] to obtain the statistical properties of the response of a statically loaded system. A system of random orthogonal polynomials is employed in the works of Iwan and Jensen [1,7] to present the response of a system with stochastic and deterministic excitation. All these papers consider related direct problems for the influence of randomly distributed parameters on the response of the system.

Herein we consider the inverse problem, i.e. the problem for stochastic identification of the properties of the system model, given the response and the excitation of the system, in the case of measured response uncertainties. The solution of the inverse problem will provide us knowledge about the influence of the uncertainties in the response on the statistical quantities of the identified parameters. In this paper we propose an approach for evaluating the statistical characteristics of the parameters to be identified, presented by their second moments, considering the response a random function of time with given probability distribution.

A lot of books and monographs are available on inverse identification problems [15,16,17]. But there is just a small number of papers concerning inverse parameter identification of systems in the case of a response with certain statistical properties [11,12,13]. The mentioned papers are aimed at the identification of the parameters of a certain model and are not applicable for other models.

It is the objective of this paper to present a quite general algorithm for inverse identification analysis in the case when the measured response is contaminated with a statistically distributed random noise. The method can be applied for parameter identification of a very wide class of systems and the only requirement is to be able to solve the direct problem either analytically or numerically.

2.Problem formulation

2.1 DETERMINISTIC FORMULATION.

Consider a system that is subjected to a given excitation (input) $U(t)$. As a result the system reacts showing a response $v_r(t)$. Suppose we have a model for the system M, which is usually expressed in the form of a system of differential equations. The model depends on certain parameters that characterize the system. Denote the parameters with the vector $\vec{\beta}, \vec{\beta}^T = (\beta_1, \beta_2, \dots, \beta_K)$, where $*^T$ stands for a transposed vector.

Then the **direct problem** is formulated as follows: Given the excitation $U(t)$ and a model $M(\vec{\beta})$ of the system. Find the response $v_r(t)$.

The related **inverse identification problem** is formulated as follows: Given the response $v_r(t)$ and the excitation $U(t)$. Find the parameter vector $\vec{\beta}$.

2.2 STOCHASTIC FORMULATION.

Suppose we can measure the response v_r and denote the measured response with $v_r^m(t)$. Assume $v_r^m(t)$ a random function of time t. Hence it can be presented in terms of its mean value plus a deviatoric component [1], as follows:

$$v_r^m(t) = v_r^0(t) + b(t) \tag{1}$$

where $v_r^0(t)$ is the expected value of v_r^m, and $b(t)$ is a zero mean process with a given autocorrelation function $S_{bb}(t_k, t_j)$, that gives the dependence between the values of $b(t)$ at different time moments. We want to determine the probability distribution of the parameter vector $\vec{\beta}$ and especially its second moment, i.e. expected value, variance and covariance.

3.Solution method.

Denote the simulated response, obtained by solving the direct problem by $v_r^s(t)$. In practice one usually gets the discrete analogs of the measured and the simulated responses, which are presented by the vectors:

$$\mathbf{v}_r^{sT} = [v_r^s(t_1), v_r^s(t_2), \ldots, v_r^s(t_n)] = (v_{r1}^s, v_{r2}^s, \ldots, v_{rn}^s)$$

$$\mathbf{v}_r^{mT} = [v_r^m(t_1), v_r^m(t_2), \ldots, v_r^m(t_n)] = (v_{r1}^m, v_{r2}^m, \ldots, v_{rn}^m)$$

where n is the number of time points.

Accordingly the representation 1 of the measured velocity becomes:

$$\mathbf{v}_r^m = \mathbf{v}_r^0 + \mathbf{b} \tag{2}$$

where $\mathbf{v}_r^0$ denotes the mean value vector of the process and $\mathbf{b}$ is a vector of random variables with zero mean and covariance matrix $\mathbf{C}$. By means of spectral decomposition of the covariance matrix $\mathbf{C}$, the random field can be presented as:

$$\mathbf{v}_r^m = \mathbf{v}_r^0 + \mathbf{F}.\mathbf{e} \tag{3}$$

or

$$\mathbf{v}_r^m = \mathbf{v}_r^0 + \sum_{l=1}^{n} \mathbf{F}_l e_l$$

where $\mathbf{F}$ is the matrix of eigenvectors of the covariance matrix $\mathbf{C}$ and $\mathbf{e}$ is a zero mean vector of uncorrelated random variables with a covariance matrix $\mathbf{\Lambda}$ given by :

$$\mathbf{\Lambda} = \mathbf{F^T C F},$$

e_l is the l-th component of the vector $\mathbf{e}$ and $\mathbf{F}_l$ is the eigenvector corresponding to the l-the eigenvalue.

The presentation 3 corresponds to the Karhunen- Loeve expansion for a discrete random function and it has been used by a number of authors [1,5,7]. It is noted that in case of a Gaussian random variable $\mathbf{v}_r^m$, the new random variables e_i are also Gaussian and since they are uncorrelated, they are independent. Subsequently the process $\mathbf{v}_r^m$ and the random field $\mathbf{e}$ will be assumed Gaussian.

The parameter vector $\vec{\beta}$ should be determined so that the simulated response $\mathbf{v}_r^s$ is as close as possible to the measured response $\mathbf{v}_r^m$. Since we are looking for the stochastic representation of $\vec{\beta}$ it is natural to present it in terms of the random vector $\mathbf{e}$. A set of orthogonal polynomials of $\mathbf{e}$ is used for the purpose:

$$\begin{aligned} \beta_i &= \mathbf{c}_i^T \mathbf{\Phi}_i(\mathbf{e}) = \sum_{j=1}^{m} c_{ij} \Phi_{ij}(\mathbf{e}), \\ &= \sum_{j=1}^{m} c_{ij} \prod_{s=1}^{n} H_{j_s}^i(e_s), \end{aligned} \tag{4}$$

where $H_{j_s}^i(e_s)$ is a polynomial of degree j_s,$j_s = 0, 1, \ldots,$, of the variable e_s, and e_s is a coordinate of $\mathbf{e}$, $s = 1, 2, \ldots, n$. The set of polynomials $\{H_l^i(e_s)\}_{l=0}^{\infty}$ is orthogonal in the random space defined by the expectation operation. That is,

$$M\left[H_l^i(e_s) H_p^i(e_s)\right] = \delta_{lp},\ H_0^i(e_s) = 1\ , \tag{5}$$

where $M[*]$ is the expectation (mean) operation. Thus the summation in 5 corresponds to a polynomial expansion of degree m in n variables. If for instance $m = 4$ the functions $\Phi_{ij}, j = 1, 2, \ldots, 4$ can be generated as follows:

$$\begin{aligned} \Phi_{i1}(\mathbf{e}) &= H_1^i(e_1) H_1^i(e_2) \ldots H_1^i(e_n) \\ \Phi_{i2}(\mathbf{e}) &= H_1^i(e_1) H_1^i(e_2) \ldots H_2^i(e_n) \\ \Phi_{i3}(\mathbf{e}) &= H_1^i(e_1) \ldots H_2^i(e_{n-1}) H_1^i(e_n) \\ \Phi_{i4}(\mathbf{e}) &= H_1^i(e_1) \ldots H_2^i(e_{n-2}) H_1^i(e_{n-1}) H_1^i(e_n) \end{aligned}$$

Since we consider Gaussian variables in this study, H_j are Hermite polynomials.

Suppose we generate a sample $\mathbf{e}_k$, $k = 1, 2, \ldots, N$ of vectors $\mathbf{e}$.(Given the distribution of the random vector $\mathbf{e}$ it is always possible to generate such a sample.) If the sample $\mathbf{e}_k$ is large and representative then eq. 5 will be satisfied by the same polynomials in the space of $\mathbf{e}_k$, $k = 1, 2, \ldots, N$.

Thus instead of determining β_i, the task is to determine the coefficients c_{ij}. Once the coefficients c_{ij} are obtained an approximation of β_i in the random space of $\mathbf{e}_k$, $k = 1, \ldots, N$ will be completely defined. For the purpose an objective function of the type below is chosen:

$$\begin{aligned} C &= \frac{1}{N} M \left\{ \left[\mathbf{v}_r^m - \mathbf{v}_r^s(\vec{\beta}) \right]^T \cdot \left[\mathbf{v}_r^m - \mathbf{v}_r^s(\vec{\beta}) \right] \right\}_{\mathbf{e}_k, k=1}^N \\ &= \sum_{k=1}^N \left[\mathbf{v}_r^m(\mathbf{e}_k) - \mathbf{v}_r^s(\vec{\beta}_k) \right]^T \cdot \left[\mathbf{v}_r^m(\mathbf{e}_k) - \mathbf{v}_r^s(\vec{\beta}_k) \right] \\ &= \sum_{k=1}^N \{ \mathbf{v}_r^{sT}(\beta_{ik}).\mathbf{v}_r^s(\beta_{ik}) + \mathbf{v}_r^{0T} \mathbf{v}_r^0 + \mathbf{e}_k^T . \mathbf{F}^T \mathbf{F} \mathbf{e}_k \\ &\quad + 2\mathbf{e}_k^T \mathbf{F}^T v_r^0 - 2\mathbf{e}_k^T \mathbf{F}^T \mathbf{v}_r^s(\beta_{ik}) - 2\mathbf{v}_r^{0T} \mathbf{v}_r^s(\beta_{ik}) \} \end{aligned} \tag{6}$$

for $i = 1, \ldots, r$, where

$$\vec{\beta}_k = \sum_{j=1}^m \mathbf{c}_j \mathbf{\Phi}_j(\mathbf{e}_k) \tag{7}$$

and accordingly

$$\beta_{ik} = \sum_{j=1}^m c_{ij} \Phi_{ij}(\mathbf{e}_k) \tag{8}$$

and

$$\Phi_{ij}(\mathbf{e}_k) = \prod_{s=1}^r H_{j_s}^i(e_{ks}) \tag{9}$$

e_{ks} being the s-th component of $\mathbf{e}_k$.

We are looking for the minimum of C with respect to c_{ij}. This means c_{ij} will be determined minimizing the squared difference between the measured and the calculated responses over all the realizations $\mathbf{e}_k$ of the random vector $\mathbf{e}$. In 6 only v_r^s depends on $\vec{\beta}_k$ and hence on c_{ij} through 7. To determine c_{ij} the following relations are employed:

$$L(c_{ij}) = \frac{\partial C}{\partial c_{ij}} = 0 \tag{10}$$

Taking into account that

$$\frac{\partial \mathbf{v}_r^s}{\partial c_{ij}} = \frac{\partial \mathbf{v}_r^s}{\partial \beta_{ik}} \Phi_{ij}(\mathbf{e}_k)$$

we obtain the following system for c_{ij}

$$L(c_{ij}) = \frac{2}{N}\sum_{k=1}^{N} \Phi_{ij}(\mathbf{e_k}) \left(\frac{\partial\, bfv_r^s}{\partial \beta_{ik}}\right) [\mathbf{v_r^s} - \mathbf{Fe_k} - \mathbf{v_r^0}] = 0 \tag{11}$$

where

$$\mathbf{v}_r^s = \mathbf{v}_r^s(\beta_{ik}),$$

β_{ik} is found from 8 and Φ_{ij} are presented according to 9.

To obtain $\mathbf{v}_r^s$ for each $\mathbf{e}_k$, the direct problem is solved with β_{ik} obtained according to 8. The derivatives $\frac{\partial \mathbf{v}_r^s}{\partial \beta_{ik}}$ are found numerically. The obtained set of equations 11 is solved iteratively. A stochastic approximation procedure [8], which leads to the next recurrence for c_{ij}:

$$c_{ij}(q+1) = c_{ij}(q) - (1/\alpha_q)\left.\frac{\partial C}{\partial c_{ij}}\right|_{c_{ij}(q)} \tag{12}$$

where q denotes the number of iterations, can be employed for the first iterations. The sequence $1/\alpha_q$ depends on the solution method applied and fulfills certain conditions [5,6]. It is advisable to switch to a faster iterative procedure after the first several iterations and this is determined for each problem separately.

Henceforward the whole procedure for obtaining c_{ij} has the following algorithm:
1)Choose an initial guess $c_{ij}(0)$;
2)Find $\beta_{ik} = \sum_{j=1}^{m} c_{ij}(0)\Phi_{ij}(\mathbf{e}_k),\ k = 1,\ldots,N$;
3) Solve the direct problem with the above β_{ik}. Obtain $\mathbf{v}_r^s(\beta_{ik})$;
4) Evaluate $\frac{\partial \mathbf{v}_{rk}^s}{\partial \beta_{ik}}$;
5) Determine $\frac{\partial C}{\partial c_{ij}}$ according to 11;
6) Find the next approximation $c_{ij}(1)$ following the iterative procedure;
7) If $|c_{ij}(0) - c_{ij}(1)| > \epsilon$ go to 2) with $c_{ij}(0) = c_{ij}(1)$, where ϵ is an error parameter.

The procedure is terminated when the difference between two sequential approximations becomes small enough, i.e. $|c_{ij}(0) - c_{ij}(1)| \le \epsilon$.

4.Statistical characteristics of the parameters

Once the unknowns c_{ij} have been determined, an approximation to the parameters in the random space is completely defined by 11 . The parameters, which are random, are described by their probability density or cumulative probability function. The second moment representation, defined by the expected values, the variances and covariances of the parameters, is of particular importance. Using presentation 5, employing relations 5, and assuming $H_j^i = H_j$ for all the parameters searched, where H_j are Hermite polynomials, it is obtained:

$$M(\beta_i) = \sum_{j=1}^{m} c_{ij} M(H_j(\mathbf{e})) = c_{i1} \tag{13}$$

$$M(\beta_i^2) = \sum_{j=1}^{m} c_{ij}^2 M(H_j^2(\mathbf{e})) + \sum_{j,k=1}^{m} c_{ij} c_{ik} M(H_j(\mathbf{e}) H_k(\mathbf{e})) = \sum_{j=1}^{m} c_{ij}^2 \tag{14}$$

$$\begin{aligned} Var(\beta_i) = \sigma_{\beta_i}^2 &= M(\beta_i^2) - [M(\beta_i)]^2 = \\ = \sum_{j=1}^{m} c_{ij}^2 - c_{i1}^2 &= \sum_{j=2}^{m} c_{ij}^2 \end{aligned} \tag{15}$$

$$M(\beta_i.\beta_k) = \sum_{j=1}^{m} c_{ij} c_{kj} \tag{16}$$

$$Cov(\beta_i, \beta_k) = M(\beta_i.\beta_k) - M(\beta_i).M(\beta_k) = \sum_{j=2}^{m} c_{ij} c_{kj} \tag{17}$$

It is noted that, in order to obtain the above formulas, it was taken into account that $\mathbf{e}$ is a Gaussian zero mean vector of uncorrelated variables, i.e.

$$M(\mathbf{e}) = 0;$$

$$M(e_i.e_j) = 0 \ .$$

5.Some examples

In the previous sections a method was proposed for the estimation of the second moment of the unknown parameters of a system with a given response, possessing stochastic uncertainties. Here the method will be applied for two completely different inverse identification problems in order to demonstrate the way it is applied and to check its performance in the special cases considered.

5.1 APPLICATION OF THE METHOD FOR THE CASE OF PARAMETER IDENTIFICATION OF ONE-DIMENSIONAL MECHANICAL MATERIAL MODELS

Consider a straight cylindrical slender and homogeneous rod specimen with density ρ. It is impacted on one end and there the excitation velocity is registered. On the opposite end the response velocity is registered. It is assumed that the longitudinal wave propagation in the specimen is governed by the following system of partial differential equations:

$$\frac{\partial \sigma}{\partial x} = \rho \frac{\partial v}{\partial t} \tag{18}$$

$$\frac{\partial \varepsilon}{\partial t} = \frac{\partial v}{\partial x} \tag{19}$$

$$\dot{\varepsilon} = \dot{\varepsilon^e} + \dot{\varepsilon^a} \tag{20}$$

$$\dot{\varepsilon^e} = \dot{\sigma}/E \tag{21}$$

$$\dot{\varepsilon^a} = G(\sigma, \varepsilon, \dot{\sigma}, \vec{\beta}...) , \tag{22}$$

where $v(x,t)$ are the point velocities in the direction of the rod, $\varepsilon(x,t)$- the strains along x,ε^e- the elastic part of the strain, ε^a - the inelastic part, $\sigma(x,t)$ - the longitudinal stresses, and $\vec{\beta}$ is a parameter vector of the material mechanical model. 18 is the equation of motion, 19 is the equation of geometry, and the last three equations 20 give the stress-strain relation that presents the material mechanical model. Here we consider the case of Maxwell-type visco-elastic models. In [6] the corresponding direct and the inverse problems are considered in details. In this work [6] we propose a deterministic parameter identification approach for the problem.

The developed method is applied for the models given in Table I with simulated responses in order to check the performance of the offered procedure and the precision it gives. The solution of the direct problem proposed in [6] is used to solve the corresponding inverse problem in the case of response uncertainties.

Table I. Material models used.

i	model M_i	G_i	β_i^T
1	Maxwell M_1	σ/η	(E,η)
2	Maxwell- Norton M_2	$\|\sigma\|^k.sgn(\sigma)/B$	(E,k,B)
3	M_3	$\sigma \exp\|\sigma\|/m/\eta$	(E,M,η)

This is done the next way:

1)The model parameter vector $\vec{\beta^i}$ (the superscript $*^i$ stands for "introduced") is introduced as a random variable with a given mean value and deviatoric component.
2)The direct problem is solved with the given $\vec{\beta^i}$ as a random variable.
3)As a result the response velocity $\mathbf{v_r^s}$ is obtained as a vector with certain statistical characteristics.
4)The measured velocity is assumed equal to $\mathbf{v_r^s}$, i.e.

$$\mathbf{v}_r^m = \mathbf{v}_r^s.$$

5)The corresponding inverse problem is solved as described in the previous sections.

6)The statistical characteristics of $\vec{\beta^o}$ (the superscript $*^o$ stands for "obtained") are obtained.
7)In order to check the performance of the developed procedure for the special case considered, the second moment of $\vec{\beta^i}$ is compared to the second moment of $\vec{\beta^o}$.

Table II. Introduced(Intr) and obtained(Obt) expectation values and standard deviations of the parameters from the verification procedure (simulated responses)

Model	Parameters	$M(*)$			$\sigma(*)/M(*)$		
		Intr	Obt	Err in %	Intr	Obt	Err in %
M_1	$E(GPa)$	3.000	3.020	0.33	0.0300	0.0302	0.66
	$\eta(GPa.s)$	12.00	11.94	0.50	0.0500	0.0504	0.80
M_1	$E(GPa)$	2.300	2.272	1.21	0.0100	0.0101	1.00
	$\eta(GPa.s)$	40.00	40.50	1.25	0.0300	0.0301	0.33
M_2	$E(GPa)$	2.500	2.478	0.88	0.0100	0.0101	1.00
	k	1.500	1.510	0.66	0.0200	0.0200	0.00
	$B(EPa^k.s)$	10.00	9.98	0.20	0.0300	0.0299	0.33
M_2	$E(GPa)$	2.400	2.420	0.83	0.0100	0.0100	0.00
	k	2.000	2.010	0.50	0.0200	0.0199	0.50
	$B(EPa^k.s)$	30.00	30.05	1.66	0.0300	0.0301	0.33
M_3	$E(GPa)$	2.000	2.020	1.00	0.0100	0.0100	0.00
	$\eta(GPa.s)$	100.0	100.9	0.90	0.0100	0.0099	1.00
	$m(TPa)$	5.000	5.020	0.40	0.0200	0.0200	0.00
M_3	$E(GPa)$	2.000	2.010	0.50	0.0100	0.0100	0.00
	$\eta(GPa.s)$	250.0	252.0	0.81	0.0300	0.0305	1.66
	$m(TPa)$	1.000	1.010	1.00	0.0300	0.0297	1.00

Table II presents the results. The relative errors in *per cent* are introduced as follows

$$e(S(\beta_i)) = \frac{|S(\beta_i^i) - S(\beta_i^o)|}{S(\beta_i^i)}.100, \tag{23}$$

where $S \equiv M$ for the case of the expected values and $S \equiv \sigma$ for the case of the standard deviations.

5.2 APPLICATION OF THE METHOD FOR THE CASE OF CRACK IDENTIFICATION FROM ELECTRIC POTENTIAL (EP) MEASUREMENTS

The developed method is applied for identifying the mean value and the variance of the parameters of a semi-elliptic crack in a rectangular prismatic body from EP measurements.

Consider a rectangular prismatic body with a surface opening semi-elliptic crack /see Fig. 1/. Constant electric potentials $\Phi = 1$ and $\Phi = -1$ are applied on the on the upper and the lower surfaces of the body, respectively. The two semi-axes of the elliptic crack and the position of its center form the vector $\vec{\beta} = (a, x, b)^T$ to be identified. It is to be identified on the basis of the measurements of the EP in

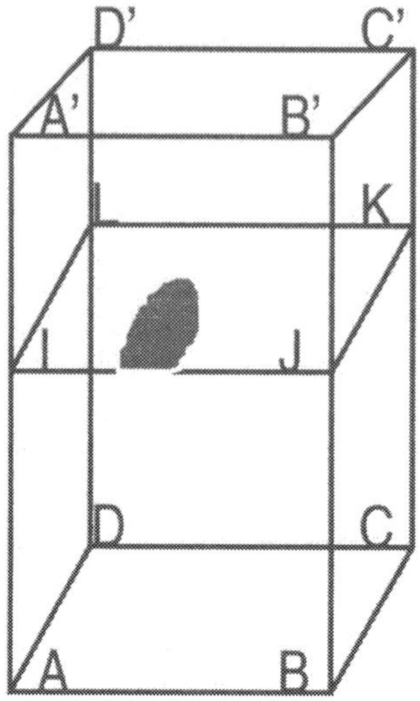

Figure 1. Rectangular prismatic body with a semi-eliptic crack.

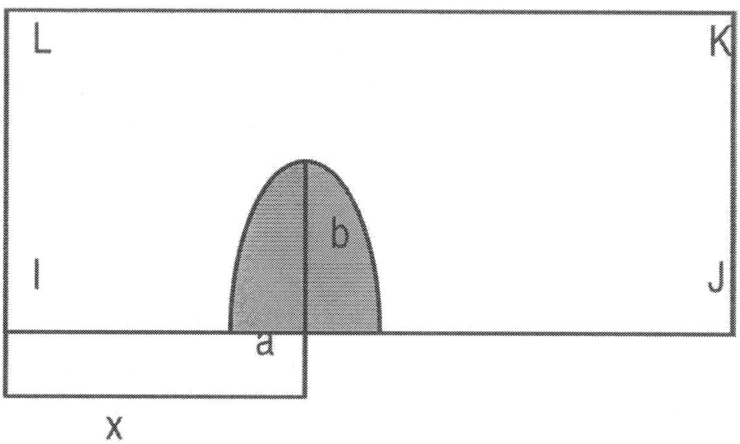

Figure 1a. Plane with semi-eliptic crack. Parameters, defining the crack, to be identified.

a number of points on the side surfaces of the prism. Thus the response vector is the vector of EP measurements **P**.

Table III. Introduced(Intr) and obtained(Obt) expectation values and standard deviations of the crack parameters (simulated responses).

Parameters	$M(*)$			$\sigma(*)/M(*)$		
	Intr	Obt	Err in %	Intr	Obt	Err in %
a	0.100	0.099	1.00	0.0100	0.0101	1.00
x	0.200	0.199	0.50	0.0100	0.0100	0.00
b	0.150	0.199	0.67	0.0100	0.0099	1.00
a	0.120	0.119	0.83	0.0300	0.0302	0.67
x	0.600	0.601	0.16	0.0300	0.0301	0.33
b	0.250	0.249	0.40	0.0300	0.0299	0.33
a	0.200	0.198	1.00	0.0200	0.0200	0.00
x	0.200	0.198	1.00	0.0200	0.0201	0.50
b	0.250	0.249	0.40	0.0200	0.0198	1.00

In this case the developed procedure for stochastic parameter identification is applied with simulated responses. This means that first the direct problem is solved introducing the parameter vector $\vec{\beta}$ as a random variable by its mean value plus a deviatoric component. As a result the response vector **P** is obtained as a random vector characterized by its mean value and a covariance matrix. The response obtained this way is fed into the inverse identification problem and finally the statistical characteristics of the vector $\vec{\beta}$ are obtained.

Table III shows the results of the method when applied for crack size and position identification from EP readings. It contains the introduced and the obtained values of the expectations and the standard deviations of the crack parameters.

6. Discussion

Herein we develop a method for inverse stochastic identification of the unknown parameters of a system in the case when the system response possesses statistically distributed uncertainties. The method is developed for a rather wide class of systems, provided the response of the system can be presented in a vector form and the solution of the corresponding direct problem is known.

The proposed method for stochastic inverse identification is developed for the case when the randomness in the response is expressed by a Gaussian random vector **b**. It can be applied for other kinds of statistical distributions, changing the type of the polynomials used in 5 the expansion of the parameter vector to be identified. It is noted that any set of linearly independent functions can be applied for the expansion 5. The use of orthogonal polynomials simplifies the calculation of the statistical characteristics of the identified parameters.

The proposed procedure was applied for two different problems with simulated responses. Tables II and III give the introduced and the obtained values of the expectations and the standard deviations of the identified parameters. The relative errors for both examples do not exceed 1%. This proves that for simulated results the proposed method is precise and adequate enough. These examples can be regarded as a verification for the developed method for the case of simulated results. This verification is done for two completely different problems and for both of them it gives consistent and precise enough results. This proves the applicability and the adequacy of the developed method for a wide class of inverse problems.

References

1. Iwan W.D, Jensen H., On the dynamic response of continuous systems including model uncertainty, *Trans ASME, J.Appl. Mech.*, vol.**25**,(1993).

2. Shinozuka M., Structural response variability, *J.Eng. Mech.*, vol.**113**, (1987),pp 825-842.

3. Yamazaki F., Shinozuka M.,Dasgupta, Neuman expansion for stochastic finite element analysis, *J.Eng. Mech.*, vol.**114**, (1988),pp 1335-1354.

4. Spanos P. and Ghanem R.,Stochastic finite element expansion for random media, *J. of the Eng. Mech. Div, ASCE*, vol.**115**, (1989),pp 1035-1053.

5. Tou J., Gonzales R., Pattern recognition principles, *Addison- Wesley Publ. Comp*, (1974).

6. Trendafilova I., Odeen S., Lundberg B., Identification of viscoelastic materials from electro-optical displacement measurements at two sections of an impacted rod specimen, *Europ. J. of Mechanics A/Solids*, **13**, No 6, (1994),pp 793-802.

7. Jensen H., W.D.Iwan, Response of systems with uncertain parameters to stochastic excitation, *Trans ASME, J. Appl. Mech,*, vol.**27**,(1995).

8. Wolfe M.A., Numerical methods for unconstrained optimization. An introduction, *Van Nostrand Reinhold C.*,(1978).

9. Hisada T., Nakagari S., Stochastic finite element method developed for strucutral safety and reliability, *3-rd International Conf. on Structural Safety and Reliability, Throndheim, Norway,*(1981).

10. Liu W.K., Mani and Belytchko T., Variational Approach to probabilistic finite elements, *J. of PEM*, vol.**2**, (1988),pp 201-213.

11. Gividini A., Maier G. and Nappi A., Parameter estimation of a static geotechnical model using a Bayes approach, *Int. J. Rock. Mech. Min. Sci*,**20(5)**, (1983),pp 215-226.

12. Maier G., Gianessi and Nappi A., Indirect identification of yield limit by mathematical programming, *Eng. Struct*, **4**, (1982),pp 86-98.

13. Nappi A., System identification for yield limits and hardening moduli in discrete elastic-plastic structures by nonlinear programming, *Appl. Math. Modeling* **6**, (1982),pp 441-448.

14. Kubo Sh., Fukumoto K., Kiyotsugu O., An application of the mollification method to estimation of governing equation of field problem from noisy observations, *submitted for publication* (1996).

15. Kubo Sh., Inverse Problems, *Baifukan* (1992).

16. Tanaka M. and H.D.Bui (Eds.), Inverse Problems in Engineering Mechanics, *Springer*,(1993) .

17. Kubo Sh.(Ed.), Inverse Problems, *Atlanta Technol. Pub*,(1993).

A COMBINED NUMERICAL/EXPERIMENTAL METHOD FOR ESTIMATING PARAMETERS AND STRUCTURE OF VISCOELASTIC CONSTITUTIVE EQUATIONS USING COMPLEX FLOWS

J.F.M. SCHOONEN , F.H.M. SWARTJES, G.W.M. PETERS,
F.P.T. BAAIJENS, H.E.H. MEIJER
Eindhoven University of Technology,
Centre for Polymers and Composites,
P.O. Box 513, 5600 MB Eindhoven, The Netherlands

Abstract: A combined numerical/experimental method is described that is used for evaluation of viscoelastic constitutive equations. The method is applied to a three dimensional stagnation flow which exhibits strong elongational deformations. The material investigated is a polymer solution (2.5% Pib/C14). Laser Doppler Anemometry and flow induced birefringece are used to measure respectively pointwise velocities and linewise (in the depth of the flow cell) stresses. Since numerical simulations of 3D viscoelastic flows are to massive yet, the deformation history of a fluid element is calculated with the viscous Carreau Yasuda model and this history is used to calculate the stresses with the viscoelastic Giesekus and PTT model. The optical birefringence signal is calculated from the integrated stresses, using the *small retardance approximation*. The approach is first tested for a slit flow (pure shear). For both models, numerical results agree well with measurements. For the stagnation flow, stresses along streamlines near the inflow and outflow symmetry plane are calculated. Using the parameters determined in simple shear flow, both viscoelastic models predict much too low integrated normal stresses in the vicinety of the outflow centreline when compared with measured values. It is demonstrated that with a proposed new constitutive model both, viscometric data and the strong elongational flow characteristics can be described at the same time.

1 Introduction

The parameters of constitutive equations for viscoelastic fluids are usually determined in simple shear only, as measurements in elongational flows are difficult to perform. Many constitutive models can fit these viscometric data equally well, though they predict different behaviour in extensional flows. It is thought that a solution to this problem is the evaluation of the constitutive equations in complex flows which are reasonably well to realize, but harder to analyse than elongational

H. Sol and C.W.J. Oomens (eds.), Material Identification Using Mixed Numerical Experimental Methods, 123-132.

flows. However, the development of reliable numerical tools make this type analyses of complex viscoelastic flows possible ((1),(2)). Besides, as these complex flows also contain (Lagrangian transient) mixed shear and elongational deformation they are even a more severe test for the constitutive equations then separate shear and elongation experiments.

The purpose of this study is the development of a combined experimental/numerical method that enables to test constitutive equations in complex flows. Figure 1 shows schematicallly our approach. Basically, the approach is the same as others used in our group to characterize anisotropic, inhomogenuous materials ((3),(4),(5)). However, in this study the residues between locally measured and predicted properties are primarily used to evaluate or even adjust constitutive models, instead of obtaining an optimal parameterset.

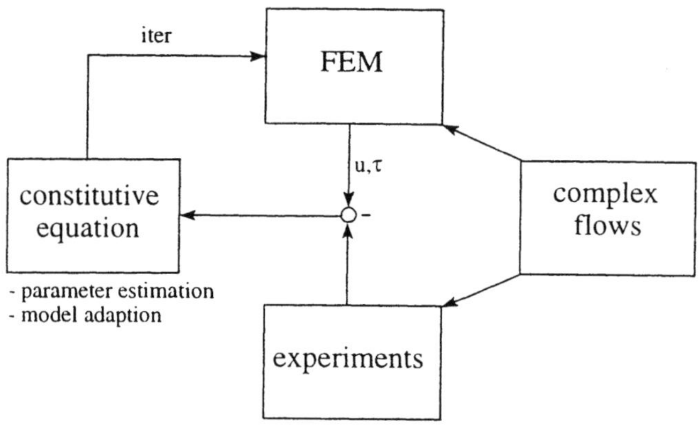

Figure 1. Schematic representation of the mixed experimental-numerical method.

The flow of a polymer solution in a stagnation die, which exhibits high elongational strains is investigated. Due to the walls of the flow cell, also shear strains will be created. Pure elongational flow is only present at the centre of the device.

Optical techniques are used to study the behavior of polymeric liquids locally. Laser Doppler anemometry (LDA) (6) to determine pointwise the velocities in the complex flow field and, following the approach of Burghardt (7), flow induced birefringence (FIB) to determine the stresses. The polymeric liquid investigated is rheologically characterized in simple shear flow first. The parameters of two nonlinear constitutive models (PTT and Giesekus) are estimated with oscillatory shear data and steady shear data. Also the parameters of a nonlinear viscous model (Carreau Yasuda) are fitted on steady shear data.

Since three dimensional finite element calculations with nonlinear viscoelastic models are to extended yet, the calculation of velocities and stresses are done separately. The flow field has been computed with the finite element package SEPRAN (8), using the Carreau Yasuda model. The velocity gradients determined are used to calculate the three dimensional stresses with the nonlinear viscoelastic models.

This approach is first tested in a fully developed rectangular slit flow (pure shear) (section 3). Then, calculations and measurements are performed on the stagnation die (section 4). In section 5 a new model is introduced and compared with measurements. Finally, the conclusions are summarized in the last section.

2 Methods and materials

2.1 Laser Doppler anemometry

Velocities were measured with a dual beam Laser Doppler anemometry system (Dantec, flow velocity analyzer 58N20), equiped with a 300 mW argon ion laser and fiber optics. The system was operated in backscatter mode, and a single velocity component was measured with the green beam ($\lambda = 514.5[nm]$). The measuring volume dimensions are ($50 \cdot 50 \cdot 200 \mu m$) and the resolution of the measured velocity field for this system is 0.13 [mm/s].

2.2 Flow induced birefringence

Flowing polymer liquids are optically anisotropic. Using the Rheo Optical Analyzer (ROA) developed by Fuller (9), the phase retardation δ and the orientation of the optic axis χ for a birefringent medium can be measured simultanuously. With the empirical stress optic rule (10), which correlates the deviatoric parts of the refractive index tensor ($\hat{\mathrm{n}}$) and the stress tensor ($\hat{\sigma}$) within a polymer flow, stresses can be calculated:

$$\hat{\mathrm{n}} = C\hat{\sigma} \tag{1}$$

where C is the stress optical coefficient. When the retardation and orientation angle are assumed constant throughout the flow cell, i.e. planar flow, the measured quantities can directly be converted into shear stress (τ_{xy}) and first normal stress difference ($N_1 = \tau_{xx} - \tau_{yy}$) (11). However, in a flow cell, three dimensional effects will always be present. Therefore the first normal stress difference and the shear stress will not be constant over the depth of the flow cell. Following Li and Burghardt (7) it can be shown that two intensities are measured for materials with a small total retardation (*small retardance approximation*).

$$R_1 = -\frac{2\pi C}{\lambda} \int N_1(z)dz \tag{2}$$

$$R_2 = \frac{4\pi C}{\lambda} \int \tau_{xy}(z)dz \tag{3}$$

This can be compared to calculations by integration of the local stresses along the depth of the flowcell (direction of the FIB lightbeam). The spatial resolution of the measuring system is 0.4 mm.

2.3 Constitutive equations

Generalized Newtonian

A generalized Newtonian model (Carreau Yasuda) (12) is used to fit the rheological steady shear data:

$$\eta = \eta_0 \left(1 + \left(\lambda_{cy}\sqrt{II_D}\right)^a\right)^{\frac{n-1}{a}} \tag{4}$$

Baaijens (11) showed that this model can accurately describe the velocity field of the flow past a cylinder of PIB/C14 solutions , indicating that, for these fluids, there is no strong influence of elongational stresses on the velocity field.

Differential viscoelastic constitutive equations

Two multimode differential constitutive equations are evaluated, both stemming from a molecular background (12). The Giesekus model is given by:

$$\lambda_i \overset{\nabla}{\tau_i} + \frac{\alpha}{G_i}(\tau_i \cdot \tau_i) + \tau_i = 2\eta_i D \tag{5}$$

and the PTT model by:

$$\overset{\nabla}{\tau_i} + \xi(D \cdot \tau_i + \tau_i \cdot D) + \frac{1}{\lambda_i}\tau_i = 2G_i D; \qquad \lambda_i = \frac{\lambda_{0i}}{1 + \frac{\epsilon}{G_i} tr(\tau_i)} \tag{6}$$

where i is the index for each mode. Giesekus has just one parameter (α), whereas PTT has two (ϵ, ξ), when all parameters are set to zero the Upper Convected Maxwell equation is recovered.

2.4 Rheological characterization in simple shear

The investigated liquid is a solution of 2.5 % PIB (Oppanol B200, BASF; $M_w \approx 4.7 \cdot 10^6[-]$) in C14 (tetradecane). Characterisation in simple shear is performed on a rotational viscometer (Rheometrics RFS-II), with a cone-plate geometry (diameter 50 [mm], cone angle 0.0199 [rad]). Both steady and dynamic measurements are performed. The Carreau Yasuda model is fitted on the viscosity curve at a reference temperature of $25°C$. Further, four Maxwell modes (η_i, λ_i) were fitted on the dynamic measurements using a Levenbergh-Marquardt algorithm. Next, the parameters of the two nonlinear viscoelastic equations were

determined using steady shear measurements, this is shown in figure 2. For the PTT model two parametersets are determined, one with parameter $\xi = 0$.

3 Steady shear flow in a rectangular channel

In this section our approach will be tested for a fully developed slit flow, where elongational forces are not present. Two channels have been investigated, which

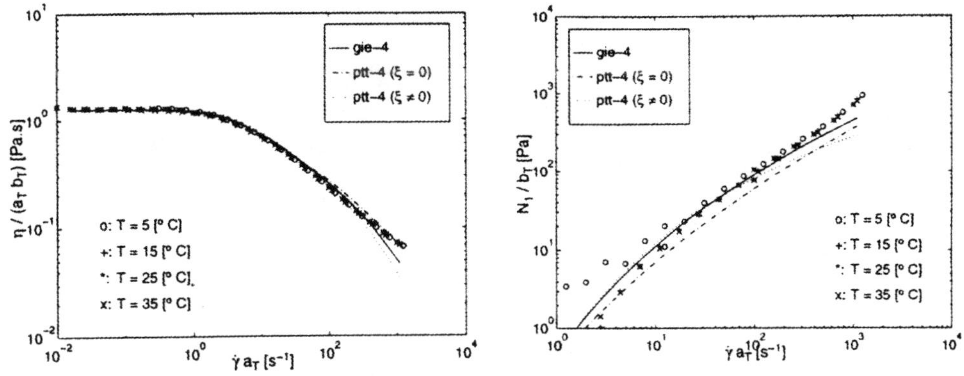

Figure 2. Characterisation in simple shear of the 2.5% Pib/C14 solution together with fits for the Giesekus and PTT model. On the left $\eta(\dot{\gamma})$, on the right $N_1(\dot{\gamma})$.

differ in their depth/height ratio (8:1 and 2:1), here only results will be presented for the small channel. The computed and measured 3D velocity profiles are shown in figure 3. Besides the velocity gradient $\dot{\gamma}_{xy}$ which in 2D flow is only present, here also $\dot{\gamma}_{xz}$ is important. As can be seen the Carreau Yasuda model accurately predicts the measured velocity field.

The 3D stresses are predicted with the viscoelastic models using the velocity gradients and are shown in figure 4. The predicted integrated stresses are shown in figure 5 together with the FIB results. The stresses also show a good agreement for all models, the largest deviation is found for the PTT model with $\xi = 0$. It can be concluded that the results for the slit flow support our approach of decoupling velocity and stress calculations in this case.

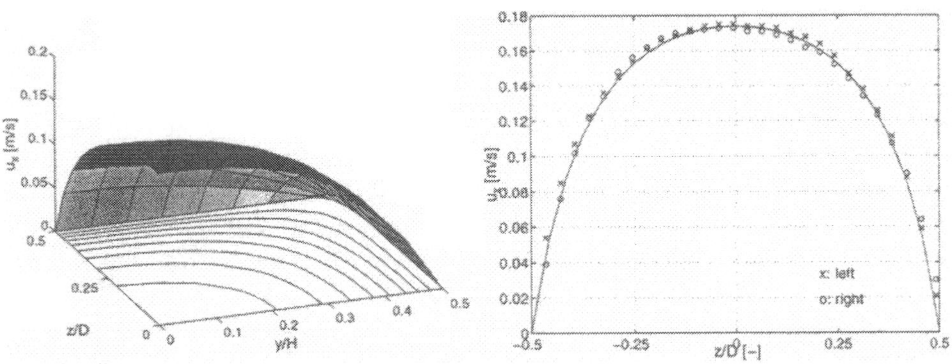

Figure 3. On the left: 3D prediction of velocity field by the Carreau Yasuda model, on the right: comparison to measured velocties along the line $y/H = 0$.

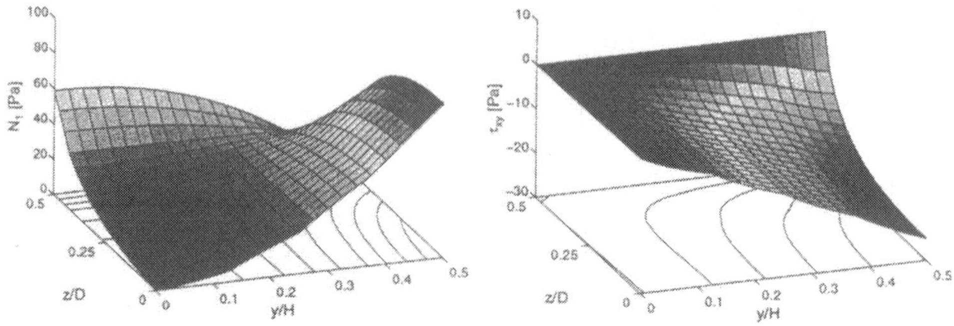

Figure 4. 3D predictions of stresses by the nonlinear viscoelastic models. On the left: first normal stress difference ($N_1 = \tau_{xx} - \tau_{yy}$), on the right: shear stress (τ_{xy}).

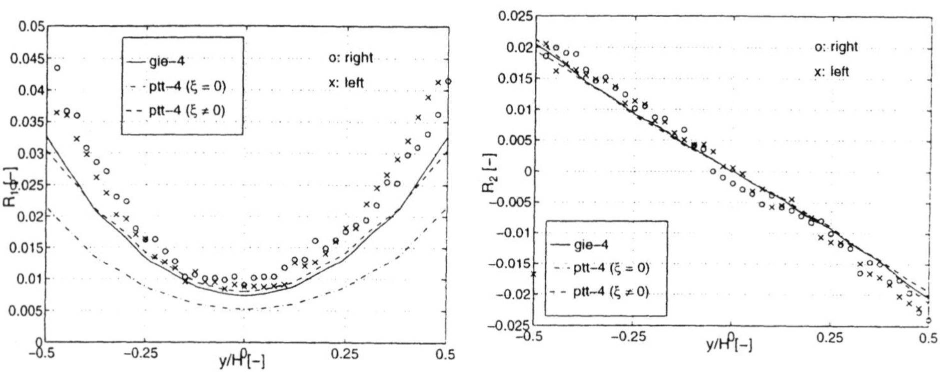

Figure 5. Comparison of measurements (intensities R_1 and R_2) with calulations (using equations 2 and 3).

4 Stagnation flow

The flow cell with its main dimensions is shown in figure 6. It consists of two perpendicular channels. In this picture the inflows are from the top and the bottom channel towards the middle of the flow cell, and the outflows are on the left and right hand side. In this way a stagnation line is created in the middle of the flow cell, causing a strong elongational flow from this line towards the outflows.
The same approach as for the slit flow is followed, the finite element mesh used to calculate the velocity field is shown in figure 6. Again, as can be seen in figure 7, using the Carreau Yasuda model gives an excellent agreement of the predicted velocities with the LDA measurements. The deformation history is determined along 24 streamlines starting from different positions along the depth at the inflow plane. Stresses are calculated along the streamlines with the nonlinear viscoelastic models using an Euler integration scheme. The integrated stresses are compared with FIB measurements in figure 8. Near the stagnation point ($x/R = 0$ in figure 8), the area where a strong elongational flow is present, all models fail to predict

the large increase in stress (R_1). This discrepancy remains present towards the outflows.

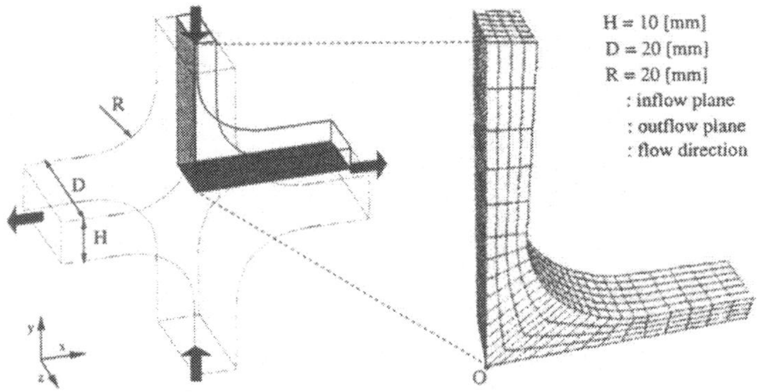

Figure 6. Schematic presentation of the flow cell together with finite element mesh.

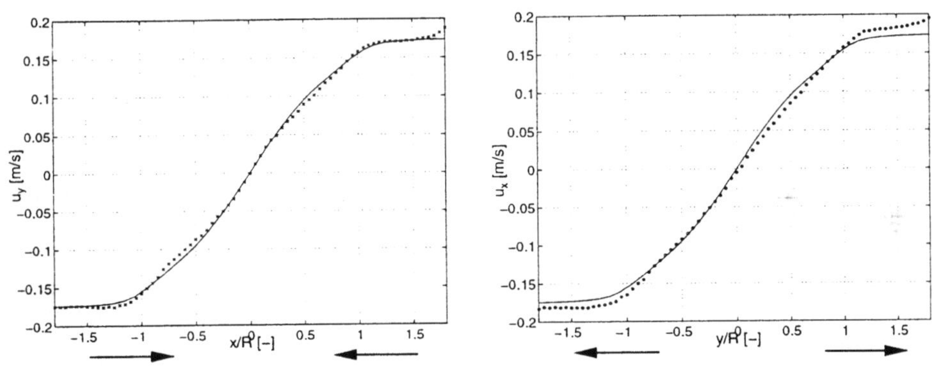

Figure 7. Comparison of the calculated velocities and LDA measurements. On the left: along the line (x,z)=(0,0) in the inflow plane as indicated in figure 6, on the right: along the line (x,y)=(0,0) in the outflow plane (figure 6).

5 New model

In the previous section it is shown that the evaluated constitutive models can not predict the stresses in the stagnation flow device with the parameters determined in simple shear flow. Changing the parameters might give a better resemblance in the elongational flow part, but on the other hand the agreement with steady shear data will become worse. Therefore constitutive equations are sought that are more flexible, preferebly with parameters that indepently control the shear and elongational behavior. An example of this, is an adjustment of the PTT model with $\xi = 0$ eq. 2.3. The structure of the model is changed by prescribing the

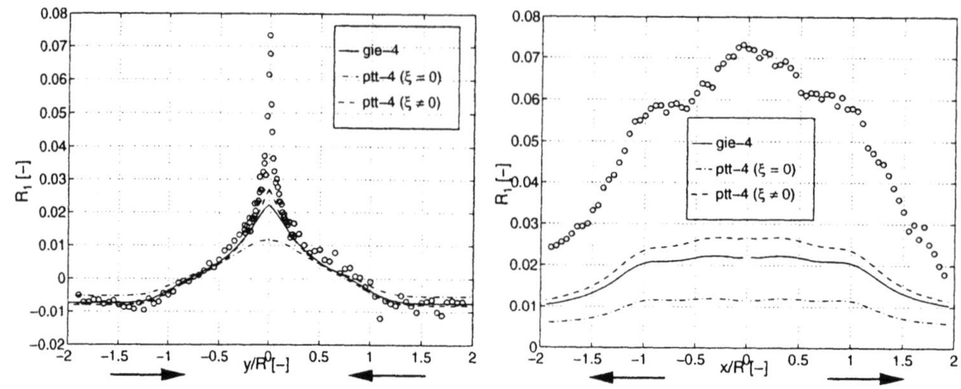

Figure 8. Comparison of the integrated stresses with FIB measurements. On the left: in the inflow plane as indicated in figure 6, on the right: in the outflow plane (figure 6).

viscosity η with an Ellis function ((13)):

$$G\lambda = \eta(\tau) = \eta_0 \left(1 + \left(\frac{II_\tau}{k}\right)^r\right)^{-1} \tag{7}$$

with λ defined in equation 2.3 In this way the viscosity is always predicted correctly, further the parameter ϵ can be adjusted for the predictions in elongational flow. When this is done for the 2.5% Pib/C14 fluid, the results for 3 different parameters ϵ are displayed in figure 9 for the slit flow, and in figure 10 for the opposed jets flow. Varying the model parameter does not change the predicted shear stress (R_2) in the slit, and the predicted first normal stress difference (R_1) only slightly changes. The predicted first normal stress difference (R_1) in the stagnation flow changes more drastically and the parameter can be adjusted to obtain an optimal fit.

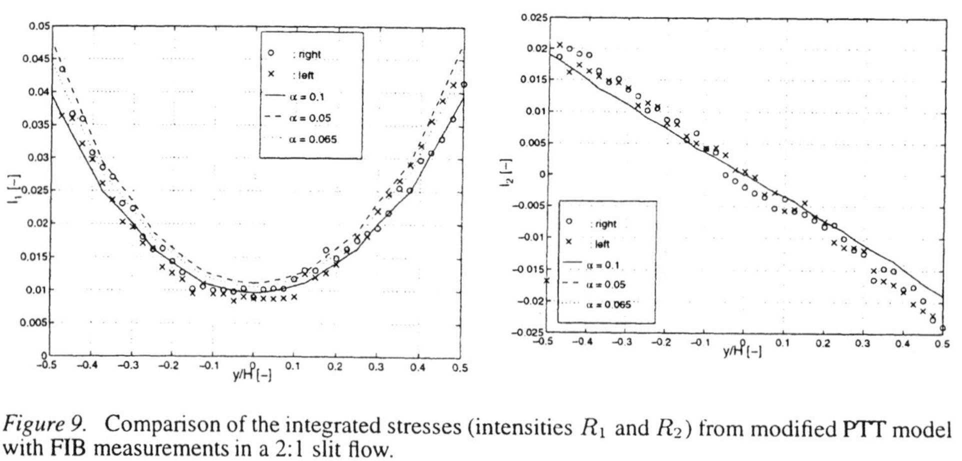

Figure 9. Comparison of the integrated stresses (intensities R_1 and R_2) from modified PTT model with FIB measurements in a 2:1 slit flow.

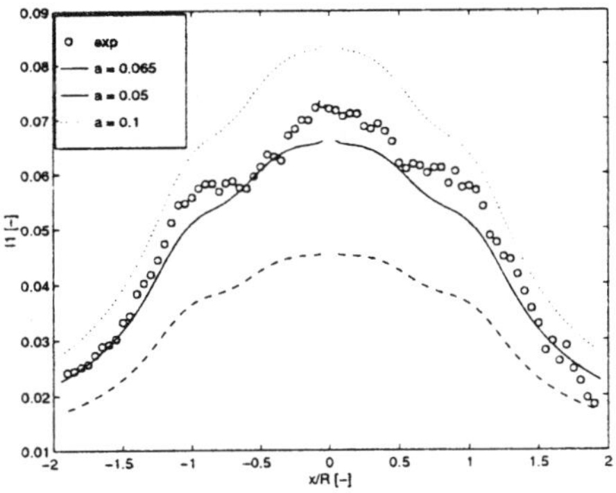

Figure 10. Comparison of the integrated stresses from modified PTT model with FIB measurements. On the left: in the inflow plane as indicated in figure 6, on the right: in the outflow plane (figure 6).

6 Conclusions

A numerical/experimental method is described which allows for the evaluation of nonlinear constitutive equations in complex flows. With the aid of optical techniques (LDA and FIB), velocities and stresses can locally be determined in the flow. It is shown that decoupling velocity (viscous model) and stress field calculation (nonlinear viscoelastic models) to do 3D flow simulations is acceptable. The evaluated Giesekus and PTT model can not predict the stresses in the stagnation flow (strong elongation) with the parameters determined in simple shear flow. With a proposed new constitutive equation, both the slit as the stagnation flow can be accurately predicted. The latter model exhibits more flexibility compared to the Giesekus and PTT model, and can almost independently adjust the material behavior in shear and elongational flow.

Acknowledgement

This work is part of a research project financially supported by the Dutch Technology Foundation (STW).

REFERENCES

[1] Baaijens F.P.T., Selen S., Baaijens F.P.T., Peters G.W.M., Meijer H.E.H. (1997) *Viscoelastic flow past a confined cylinder of a low density polyethylene melt*, Journal of Non-Newtonian Fluid Mechanics, **68**, 173-203.

[2] Peters G.W.M., Baaijens F.P.T. (1997) *Modelling of non-isothermal viscoelastic flows*, Journal of Non-Newtonian Fluid Mechanics, **68**, 205-224.

[3] Ratingen van M.R. (1994) *Mechanical identification of inhomogeneous Solids, a mixed numerical experimental approach*,PhD thesis, Eindhoven University of Technology, Eindhoven, the Netherlands.

[4] Meurs P.F.M., Schreurs P.J.G., Peijs T., Meijer H.E.H. (1996) *Characterization of Interphase Conditions in Composite Materials*, Composites **27A**,

781-786.

[5] Meeuwissen M.H.H, Oomens C.W.J., Baaijens F.P.T., Petterson R., Janssen J.D. (1997) *Determination of parameters in elasto-plastic models of aluminium*, in: Material identification using mixed numerical experimental methods (EUROMECH 357), Kerkrade, the Netherlands.

[6] Drain L.E. (1980) *The Laser Doppler Technique*, John Wiley and Sons, Chicester.

[7] Li J.-M., Burghardt W.R. (1995) *Flow birefringence in axisymmetric geometries*, Journal of Rheology **39**, 743-766.

[8] Segal A. (1992) *Manual SEPRAN package*, Ingenieursburo SEPRA, Leidschendam, The Netherlands.

[9] Fuller G.G., Mikkelsen K.J. (1989) *Note: Optical rheometry using a rotary polarization modulator*, Journal of Rheology **33**, 761-769.

[10] Janeschitz-Kriegl H. (1983) *Polymer melt rheology and flow birefringence*, volume 6 of *Polymers/Properties and Applications*, Springer-Verlag.

[11] Baaijens H. (1994) *Evaluation of constitutive equations for polymer melts and solutions in complex flows*, PhD thesis, Eindhoven University of Technology, Eindhoven, the Netherlands.

[12] Larson R.G. (1988) *Constitutive equations for polymer melts and solutions*, Buttersworth, Boston.

[13] Macosko C.W. (1994) *Rheology, principles, measurements, and applications*, VCH Publicers, New York.

COMPUTER AIDED IDENTIFICATION OF THE TEMPERATURE DEPENDENCE OF THERMAL PROPERTIES

P. KOC and B. ŠTOK
Faculty of Mechanical Engineering, University of Ljubljana
Aškerčeva 6, Ljubljana
SLOVENIA

The modern material science is largely engaged in processing of advanced materials. Since the behaviour of such materials is usually characterized by complex relationships an efficient material characterization is demanded, along with the material processing. Accordingly, special experimental techniques and material identification methods are developed. This paper considers the characterization of thermal properties of an anisotropic solid, as determined by a numerical solution of the corresponding inverse problem. In particular, a transient heat transfer in an orthotropic cylindrical specimen is considered.

1. Introduction

It has been long recognized that materials can be properly characterized only by adequate laboratory experiments. Often, due to complex material behaviour different specific experiments, limited usually to simple loading cases and specimen geometries, are needed for a proper characterization of the same material. Usually, the laboratory experiments used for the corresponding material characterization are designed in such a way, that only one or two material properties can be identified with one experiment. Consequently, several experiments of different kind have to be performed in order to obtain complete material characterization. Not always, however, the complete characterization is achieved. In contrast to this experimental experience, any computer based approach that is aiming at loosening of the severity of conditions imposed in an experiment is most welcome. Such approach would hopefully lead to possible simplification of the experimental set-up itself and to reduction of individual experiments needed for the complete material characterization.

In the evaluation of the experimental data the computer assistance has been present for a rather long time, its role in the material characterization being most significant. Nowadays, when considering recent advances in computational techniques for the solution of inverse problems, this assistance becomes even more firm and efficient. By introducing computer simulations to resimulate a real complex experiment, a powerful means for the material characterization is established. Namely, when performing a computer simulation under the same conditions, as met in a real experiment, it is possible, by imposing the equivalence of the computed and measured responses, to identify material properties.

H. Sol and C.W.J. Oomens (eds.), Material Identification Using Mixed Numerical Experimental Methods, 133-141.

2. Problem definition

In technical applications we often meet problems that are characterized either by transient or steady state nonuniform temperature fields. The temperature dependence of various physical properties is of particular interest in such problems. A method for the identification of physical properties that influence the heat transfer in a solid body will be presented in the sequel. The temperature field $T=T(x_k,t)$ in a solid that is subjected to transient heat conduction is governed by the following differential equation in the tensorial form

$$\varrho c \frac{\partial T}{\partial t} = (k_{ij} T_{,i})_{,j} + Q \tag{1}$$

and associated initial and boundary conditions. Here, x_k and t represent the space and time coordinate, respectively, while Q stays for the heat sources. Physical properties that characterize the heat conduction are specific density ρ, specific heat capacity c and thermal conductivity k_{ij}, all of them being temperature dependent. By nature being a symmetric tensor the thermal conductivity in an anisotropic body is direction dependent, its structure being determined by a number of independent non-zero coefficients. With respect to the coordinate system used, i.e. an arbitrary one or the one defined by the material principal directions, this number is 6 or 3, respectively. This number reduces to one single non-zero coefficient in the case of material isotropy [3].

Analysing the structure of Eqn. (1) in the sense of material identification, where ρ and c are appearing in a product form, it can be concluded that a simultaneous and unique identification of ρ and c is not possible. Correspondingly, either the product of both properties or, assuming one property being known, the other property can be identified. Now, we assume the temperature dependence of specific density $\rho=\rho(T)$ as known, and accordingly, only the temperature dependence of specific density $c=c(T)$ and thermal conductivity tensor $k_{ij}=k_{ij}(T)$ will be identified.

A computer aided identification of the temperature dependence of any thermal property is based on an assumed functional structure $F=F(T)$ with temperature T as independent variable. Any mathematical form of a continuous function F can be assumed in principle, but the degree of agreement with the real material behaviour depends on the number of free parameters in it. In the considered case, the sectionally linear approximation of function F is assumed, and accordingly, the temperature domain of interest $[T_s, T_f]$ is divided into m number of intervals

$$[T_s, T_f] = [T_0, T_1] \cup (T_1, T_2] \cup \ldots \cup (T_{m-1}, T_m] \quad , \quad T_s \equiv T_0 \quad , \quad T_f \equiv T_m \tag{2}$$

In accordance with the assumed function approximation the value of $F(T)$ at specific temperature $T \in [T_{i-1}, T_i]$ can be expressed as

$$F(T) = F^{i-1} + \frac{F^i - F^{i-1}}{T_i - T_{i-1}} (T - T_{i-1}) \quad , \quad 1 \leq i \leq m \tag{3}$$

where coefficients F^{i-1} and F^i are the corresponding values of function $F(T)$ at the borders of the i-th interval.

In the material characterization process the parameters of the specific heat capacity function $F(T) \equiv c(T)$ and thermal conductivity function $F(T) \equiv k_{ij}(T)$ are material constants which are to be denoted, in the sense of Eqns. (2) and (3), as c^0, c^1,..., c^a and $k_{ij}^{\,0}$, $k_{ij}^{\,1}$,..., $k_{ij}^{\,b}$, respectively. Here, a and b represent the corresponding numbers of intervals for the respective function approximations. The number of thermal conductivity coefficients $k_{ij}^{\,r}$, $0 \le r \le b$, to be considered in the case of material anisotropy depend on the coordinate system used. For an arbitrary coordinate system the corresponding tensor structure is of the form

$$k_{ij}^{\,r} = \begin{bmatrix} k_{xx} & k_{xy} & k_{xz} \\ k_{yx} & k_{yy} & k_{yz} \\ k_{zx} & k_{zy} & k_{zz} \end{bmatrix}^r \quad , \qquad k_{ij} = k_{ji} \tag{4}$$

whereas for the coordinate system, identical to the principal material directions, this form reads

$$k_{ij}^{\,r} = \begin{bmatrix} k_1 & 0 & 0 \\ 0 & k_2 & 0 \\ 0 & 0 & k_3 \end{bmatrix}^r \tag{5}$$

Rotation of the latter, i.e. the orthotropic material coordinate system, with respect to the global arbitrary one is given in the unique way by three independent Euler's angles φ, ψ and χ (Fig. 1).

For purposes of further mathematical manipulation function coefficients c^r and k_{ij}^r are assembled in a material parameters vector p

$$p^T = [p_1,...,p_n] = [c^0,...,c^a,k_{ij}^0,...,k_{ij}^b] \tag{6}$$

where n is the total number of coefficients to be identified.

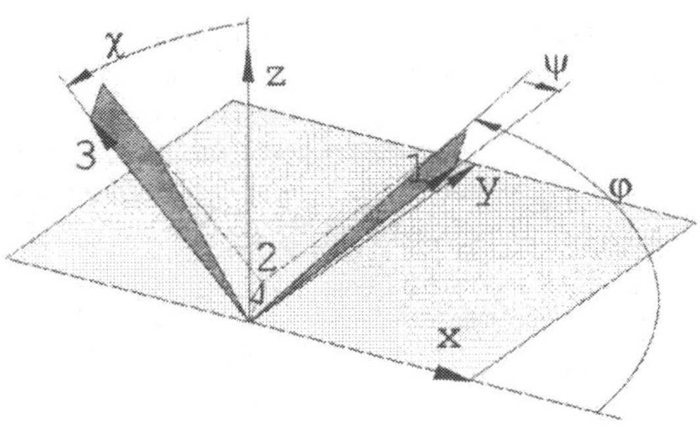

Figure 1. Principal conductivity directions

3. Method of parameter estimation

A relationship that governs the response of a system, which is subjected to applied actions, can be in general described by operational equation of the form

$$Ax = f \tag{7}$$

When the structure of operator A is not known the problem of material characterization consists first of the identification of the governing rule, and afterwards, the structure of operator A being identified, it is followed by the estimation of material parameters that determine the system behaviour quantitatively. Since in our case the governing rule is adopted, Eqn. (1), the task is to estimate the unknown material parameters assembled in vector p, and the problem can be classified as an inverse problem of the second kind.

In the problem of material characterization a solution of the inverse problem is actually following solution strategies used in extremum problems [1]. In the identification / estimation problems responses of two systems are considered, one being related to a real physical model and the other being related to a mathematical model which is supposed to represent the behaviour of the physical model equivalently. The objective of any such problem is to enforce the response of the assumed mathematical model to be as close as possible to the physical one. Of course, this is achieved only by a proper adjustment of material parameters, which finally yields the sought material characterization. In the sequel, when referring to the considered two systems, the associated quantities will be addressed as "experimental" or "numerical".

The scalar objective function $S(p)$, which is to be minimized, is written as a matrix product

$$S(p) = [Y - U(p)]^T W [Y - U(p)] \tag{8}$$

where Y and $U(p)$ are vectors formed from a subset of the corresponding "experimental" and "numerical" quantities. The role of weighting matrix W is to introduce unequal weights due to measurement errors and due to physically different quantities. Superscript T indicates the matrix transpose. The computed response $U(p)$, which is usually obtained by the Finite Element Method analysis, is a function of material parameters vector p, assumed in the computational analysis. No regularization term, which includes vector p in an explicit manner in the objective function, is needed.

In this work, a method used for the minimization of Eqn. (8) is the Gauss-Newton method of minimization [2]. Although more powerful methods of minimization exist, particularly when referring to the structural optimization domain [1], it seems that they are less efficient for identification purposes.

The necessary condition for $S(p)$ to be stationary is that the matrix derivative of $S(p)$ with respect to p be equal to zero

$$\nabla_p S(p) = -2 X^T(p) W [Y - U(p)] = 0 \tag{9}$$

where the sensitivity matrix $X(p)$ is defined as

$$X(p) = \left[\nabla_p U^T(p)\right]^T \tag{10}$$

Vector p, that minimizes the considered objective function in accordance with Eqn. (9), is obtained iteratively. With ν being the iteration counter the relationship relating the responses U^{ν} and $U^{\nu+1}$ in two subsequent iteration computations can be approximated by the linearized Taylor series expansion of U in the parameter space

$$U(p^{\nu+1}) = U(p^{\nu}) + X(p^{\nu})[p^{\nu+1} - p^{\nu}] \tag{11}$$

The substitution of this relationship in Eqn. (9) yields a set of equations

$$X^T(p^{\nu})\, W X(p^{\nu}) p^{\nu+1} = X^T(p^{\nu})\, W\left[X(p^{\nu})p^{\nu} + [Y - U(p)]\right] \tag{12}$$

which is used as the iterative solution algorithm for obtaining a new estimate of vector p.

The search procedure for finding the extremum of objective function $S(p)$ is relying on the sensitivity analysis of each single parameter in vector p, establishing thus its influence on the calculated primary variables. Due to high implicit dependence and nonlinearity the solution is obtained iteratively. The sensitivity coefficients are calculated using finite differences, requiring thus in an n-parameter problem $(n+1)$ direct problem analyses for each iteration. The finite difference approach of the sensitivity analysis being very time consuming it should be replaced with analytical or semi-analytical methods, whenever possible.

4. Numerical example

To illustrate the proposed method of inverse material identification an example of heat conduction in an anisotropic solid body is considered. Temperature dependent specific heat capacity and thermal conductivity tensor are sought in a wide temperature range. The specimen is made of composite material (graphite fibre +epoxy) with material properties given in [4].

4.1. PHYSICAL EXPERIMENT ARRANGEMENT

The experimental set-up, which is at the beginning of the experiment at uniform temperature T_0, consists of a heater, a specimen, temperature sensors and the corresponding insulation (Fig. 2). The specimen's geometry used for the thermal properties identification depends on whether or not at least one material principal direction of an investigated anisotropic material is known. If this direction is known a concentric annular plate sample, with the rotational axis being identical to the known principal direction, is used. Otherwise, a concentric hollow cylinder sample must be used. From the heater the heat flux q of known, i.e. measurable, magnitude is applied on the inner surface of the specimen. It can act either on the whole surface or just on a part of it, but always in such a way that rotational symmetry of boundary conditions is preserved. All surfaces of the specimen, except the heated ones, are insulated. Since the sensors for temperature measurement are fastened at the accessible surfaces of the specimen no additional drills for sensors are needed. The sensors are attached on the cylindrical surfaces of the specimen, concentrated at locations where significant temperature variations are expected. In order to keep the number of sensors as low as possible the proper choise of sensor locations is of great importance.

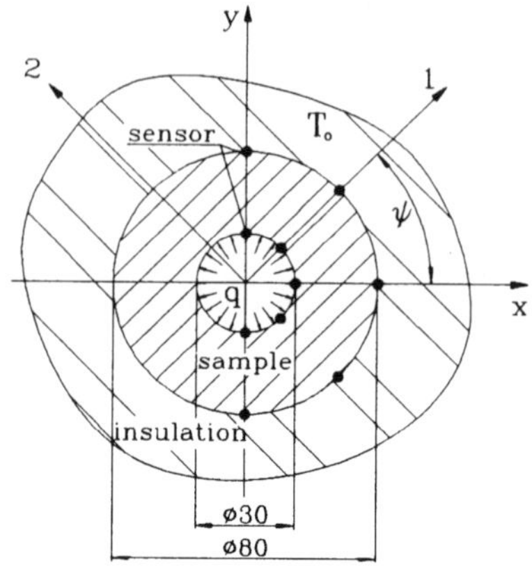

Figure 2. Experiment arrangement

4.2. EQUIVALENT NUMERICAL EXPERIMENT

Temperatures that are measured at the sensor locations represent the response of a physical experiment, in Eqn. (8) denoted by Y. For the investigation purposes only the material characterization, following the previously described approach, will be performed on an equivalent numerical experiment that will replace the physical one. Accordingly, the measured data in a physical experiment will be substituted by the computed "experimental" response, as provided by the direct FEM analysis of the observed system when assuming the material properties known. It is clear that, in comparison to the physical experiment, the corresponding numerical model is only an idealization, and consequently, it can not take defficiencies and imperfections that are characterizing any real experimentation into account.

The thermal properties, listed in Table 1 as specified in [4], are the basis for the computed "experimental" response. At the same time they are the target values when performing the inverse identification.

TABLE 1. Thermal properties of composite material

T [°C]	c [J/kgK]	k_1 [W/mK]	k_2 [W/mK]	k_3 [W/mK]
-173	337	5.7	0.46	1.0
-73	642	8.7	0.68	1.0
27	935	11.1	0.87	1.0
127	1216	13.0	1.10	1.0

The specific density, which is not a subject of the identification, has a constant value of 1400kg/m^3.

The numerical experiment is conducted for the case of planar anisotropy with a flat specimen having the axis of rotation (z-axis) identical to the material principal axis. In the specimen plane the principal material axes (1,2) are rotated against the global coordinate system (x,y) for an angle of ψ=45°. The heat flux of magnitude q=60 000W/m^2 is imposed on the inner cylindrical surface of the specimen in the total duration of 90 seconds.

From the computed "experimental" response results a subset of temperatures is formed to provide the "measured" data set, which is to be used subsequently in the material parameters estimation. Actually, we assume as "measured" the temperatures of surface points of the FE model which coincide with the sensors' locations. To thus obtained "experimental" data an artificial measurement error is added. Usually, conducting an experiment any known systematic effects are assumed to be removed, so that the remaining errors can be considered as random and their effect can be described statistically. The random error associated with temperature measurements can be quantified as

$$T_{err} = R \cdot \max[2.5°C,\ 0.0075\Delta T] \quad , \qquad -1 \leq R \leq 1 \tag{13}$$

where ΔT is the temperature difference between the warm and cold thermocouple junctions, and R is a Gaussian distributed random number.

4.3. RESULTS OF IDENTIFICATION

Thermal properties that are to be characterized are identified by solving the corresponding inverse problem, where in each iteration ν a set of equations, Eqn. (12), is to be solved. The computed response $U(p^\nu)$ is obtained by solving a direct heat conduction problem, Eqn. (1), with the material behaviour defined in accordance with the actual value of the material parameters vector p^ν. The numerical model is analysed by the same FEM mesh as it was used in the analysis that enabled the acquisition of a computed "experimental" response, though there is no special reason for it. In a real experiment, the mesh accuracy test should be performed in any case. In order to avoid eventual influence on the inverse identification, due to the possible similarity or even following of the same computational path, the time incrementation is not the same for the two analyses. So, a variable time step length scheme is applied.

The identification of the considered orthotropic material is elaborated twice, first by assuming its general form in the (x,y) plane, Eqn. (4), and second by assuming its specific form in the space of principal material directions, Eqn. (5). When considering the first case 3 planar conductivity coefficients, k_{xx}, k_{yy} and k_{xy}, respectively, are of interest, their identification being related to the actual temperature domain and corresponding intervals. Transforming the conductivity tensor from the global coordinate system (x,y,z) to the principal direction coordinate system (1,2,3), principal conductivities k_1 and k_2, and principal direction vectors are obtained, the latter being uniquely defined by angle ψ.

Assuming a planar anisotropic material in the (x,y) plane and identifying the considered thermal properties in the range of temperatures, as given in Table 1, the material parameters vector p consists of 16 independent parameters

$$p^T = [c^0,\dots,c^3,k_{xx}^0,\dots,k_{xx}^3,k_{yy}^0,\dots,k_{yy}^3,k_{xy}^0,\dots,k_{xy}^3] \tag{14}$$

The initial estimate for the parameters assumes that they are constant on the whole temperature range of interest

$$
\begin{aligned}
c^r(T) &= 500 J/kgK \\
k_{ii}^r(T) &= 10 W/mK \quad , \qquad r = 0,\dots,3 \\
k_{ij}^r(T) &= 5 W/mK \quad , \qquad i \neq j \; ; \;\; i,j \in \{x,y\}
\end{aligned}
\tag{15}
$$

The results of the identification were obtained after 13 iterations and are tabulated in Table 2. As it can be seen by comparing the obtained results with the ones in Table 1 the quality of the identified material parameters is pretty good. The fact that principal material directions do not alter during heating, i.e. angle ψ remains unchanged, can be used for further quality validation test. The agreement with the given angle of ψ=45° is excellent.

TABLE 2. Identified parameters - planar anisotropic conductivity tensor

	Identified values				Transformed values		
T [°C]	c [J/kgK]	k_{xx} [W/mK]	k_{yy} [W/mK]	k_{xy} [W/mK]	k_1 [W/mK]	k_2 [W/mK]	ψ [deg.]
-173	338	3.123	3.114	2.666	5.8	0.45	44.95
-73	642	4.658	4.674	3.978	8.6	0.69	45.05
27	912	6.092	6.065	5.197	11.3	0.88	44.92
127	1300	7.233	7.334	6.345	13.6	0.94	45.23

In the case of planar orthotropy the material parameters vector p consists of 13 independent parameters, given in Eqn. 16:

$$
p^T = [c^0,\dots,c^3,k_1^0,\dots,k_1^3,k_2^0,\dots,k_2^3,\psi]
\tag{16}
$$

where ψ is the angle of rotation between x-axis and 1-coordinate axis (Fig. 2). Here, we take into account that principal material directions do not alter with temperature variation, i.e. angle ψ is constant throughout.

The results of identification, the initial estimate for parameters being given again by Eqn. (15), are obtained after 20 iterations. Their tabulated values are listed in Table 3 while the identified angle ψ has reached a value of 44.99°. Again, the identification is very efficient and in good agreement with the assumed material properties.

TABLE 3. Identified parameters - planar orthotropic conductivity tensor

T [°C]	c [J/kgK]	k_1 [W/mK]	k_2 [W/mK]
-173	343	5.8	0.47
-73	634	8.7	0.68
27	958	11.0	0.84
127	1270	12.6	1.16

5. Conclusion

In the paper an identification methodology used for the determination of thermal properties of materials that exhibit anisotropic behaviour is presented. The method can be used generally for the estimation of an arbitrary number of parameters. However, because of the excessive time consuming identifications when estimating a larger number of material properties, the number of parameters is limited. The investigation of planar anisotropy has shown that in both cases of anisotropic descriptions, orthotropic as well as general, the inverse identification of material parameters is efficient and reliable.

The method can be used as a tool for the optimal design of experiments. Relying on the sensitivity analysis, obtained from the consideration of the numerical model and the corresponding simulation of a real experiment, the points in time and space domain with the best data acquisition possibilities can be obtained.

6. References

1. Arora J.S.: *Introduction to Optimum Design*, McGraw-Hill Book Company, New York, 1989.
2. Beck J.V., Arnold K.J.: *Parameter Estimation in Engineering and Science*, John Wiley & Sons Inc, New York, 1977.
3. Padovan J.: Anisotropic Thermal Stress Analysis in Hetnarski B.R. (ed.) *Thermal Stresses, Vol. 1*, Elsevier Science Pulishers, Amsterdam, 1986, pp. 143-262.
4. Incropera F.P. and DeWitt D.P.: *Fundamentals of Heat and Mass Transfer*, John Wiley & Sons Inc., New York, 1987.

COMPLEX MATERIAL PROPERTY IDENTIFICATION FOR CEMENT MATRIX COMPOSITES BY A MIXED NUMERICAL-EXPERIMENTAL METHOD

J. VANTOMME[1], J.M. NDAMBI[1], J. DE VISSCHER[2], H. SOL[2], W.P. DE WILDE[2]

1 *Civil Engineering Department, Royal Military Academy, Brussels, Belgium*

2 *Department of Structural Analysis, Faculty of Engineering, Free University of Brussels VUB, Belgium*

Abstract

The paper uses a mixed numerical-experimental technique (MNET) to determine material properties of cement matrix composites. The MNET is based on modal analysis of free vibrating plates, and is applied for the case of microconcrete and polymer modified cement matrix. The obtained values for the complex moduli are discussed, as well as the influencing parameters. It is concluded that both stiffness and damping appear to be sensitive parameters for the characterization of concrete materials. It is also noted that the methodology and exploitation of the MNET is somewhat different for concrete than for polymer composite materials, for which the MNET was originally developed.

Next, the MNET is also used for the full characterization of reinforced cement matrix composites. Attention is focused on the theoretical moduli predictions for reinforcement under the form of randomly oriented steel fibres. These models are based on the energy balance approach, which subdivides the global strain energy into basic portions, each related to a fundamental stress-strain relationship. It is again shown that earlier models for fibre reinforced plastics cannot be simply transposed to the case of cement matrix composites, in spite of the common basic composite concept.

H. Sol and C.W.J. Oomens (eds.), Material Identification Using Mixed Numerical Experimental Methods, 143-152.

1. Introduction

Single-property experiments to determine elastic material parameters often yield poor results for composite concrete systems, especially for the transverse and shear properties. This paper presents a different approach to establish these properties by means of experimental information taken from a structural element, which is linked to the numerical output by a finite element model [1]. The paper first briefly presents the basic idea of the identification procedure, which is then applied onto a standard epoxy mortar and a polymer modified cementitious repair mortar, in order to show that full material characterization can be achieved. Next, models for reinforced systems are considered.

The characterization starts from the basic assumption that the mortars may be considered as macroscopically homogeneous, continuous and isotropic materials, whose mechanical behaviour may be described according to the general linear viscoelastic model. A convenient way of describing viscoelastic material behaviour is the complex notation, introducing the concept of complex modulus, which is in general physically explained by considering the phase difference δ between sinusoidally varying stress σ^* and strain ε^* for uniaxial loading:

$$\frac{\sigma^*}{\varepsilon^*} = \frac{\sigma_0\, e^{i\omega t}}{\varepsilon_0\, e^{i(\omega t-\delta)}} = \frac{\sigma_0}{\varepsilon_0}\, e^{i\delta} = K' + i\,K'' = K^* \tag{1}$$

with : - the storage modulus K' representing the elastic behaviour;
- the loss modulus K", representing the energy dissipation;
- the angular frequency ω.

According to the linear viscoelastic model (or Kelvin-Voigt model), these two moduli can also be related to damping characteristics:

$$\frac{K''}{K'} = \operatorname{tg}\delta = \eta \tag{2}$$

with δ the loss angle and η the loss factor.

2. Determination of the elastic moduli

The identification method was practically set up by Sol [2] for the case of orthotropic materials. The passage to isotropic materials is straightforward.

The procedure starts with the measurement of resonance frequencies of plate specimens and the extraction (partially based on an empirical relationship) of the plate rigidities, which are related to the elastic moduli: expressions (3).

$$D_{11} = \frac{t^3}{12} \frac{E_1}{1-\nu_{12}\nu_{21}} \qquad D_{22} = \frac{t^3}{12} \frac{E_2}{1-\nu_{12}\nu_{21}}$$
$$D_{12} = \nu_{21} D_{11} \qquad D_{66} = \frac{t^3}{12} G_{12} \tag{3}$$

where: - t is the plate thickness;
- 1, 2 are the principal material directions.

The plate rigidities are iteratively updated until the numerically computed resonance frequencies completely match the measured frequencies.

As to the numerical model, Sol consideres orthotropic thin plates, subjected to small deflections; the plate may be modelled with classical laminated plate theory, which is based on Love-Kirchhoff assumptions. Starting from the equilibrium equations of an orthotropic plate and applying the Rayleigh-Ritz technique, he finds a generalised eigenvalue problem, the numerical solution of which yields the eigenvalues and mode shapes, which are compared with the experimentally measured values. If the difference is not sufficiently small, a new estimation for the plate rigidities is computed by means of a parameter estimation method. Combining numerical and experimental techniques, a parameter estimation method will adapt the parameters in the numerical model such that the newly calculated eigenfrequencies (system analysis) match the experimental values (system identification).

3. Determination of damping properties

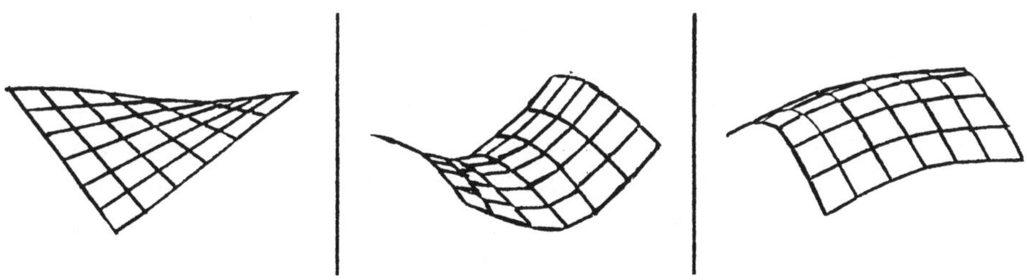

Figure 1: Torsion (T), Saddle (S) and Breath (B) mode shapes.

The previously cited method to determine the elastic material properties or storage moduli, has been extended to the determination of complex moduli.

Taking into account the above mentioned assumptions concerning mortar material, the test structure is a square plate with free boundary conditions, whose dimensions are chosen in order to satisfy the Love-Kirchhoff assumptions, thus involving a plane stress condition. The first three modes of the plate are in order of increasing resonance frequency: the Torsion (T), Saddle (S) and Breath (B) mode; the mode shapes are represented in figure 1. As reported in [3], the modal loss factor may be composed of four terms:

$$\eta_{mode} = \eta_{D_{11}} \frac{W_{11}}{W} + \eta_{D_{22}} \frac{W_{22}}{W} + \eta_{D_{12}} \frac{W_{12}}{W} + \eta_{D_{66}} \frac{W_{66}}{W} \qquad (4)$$

with : - η_{Dij} the loss factor associated with the plate rigidity D_{ij};
- W_{ij}/W the fraction of the total strain energy W for the considered vibration mode, associated with the plate rigidity D_{ij};
- $\eta_{D11} = \eta_{D22}$ due to the assumption of isotropy.

The fractions of elastic strain energy W_{ij}/W may be calculated for the T, S and B mode, by means of a finite element model, based on the determined elastic properties, while the three modal loss factors η_T, η_S and η_B can be measured experimentally. Assuming that η_{Dij} is independent of frequency and rewriting expression (4) for each mode, turn out to be a system of three equations in three unknowns: $\eta_{D11} = \eta_{D22}$, η_{D12} and η_{D66}. Once determined, these three loss factors are then used to determine the loss factors η_E, η_G and η_v, by exploitation of the expressions (3), which are written in complex form.

4. Practical measurement procedure

In the case of isotropic materials, testing is performed on a square plate. In the case of orthotropic materials (mortars with continuous reinforcement), the dimensions of the plate should be as such to verify certain conditions [2], in order to avoid a degeneration of the T, S and B mode shapes.

The plate is suspended from a frame and is excited by means of a hammer. The input signal is captured by a force transducer, fixed on the hammer. Although accoustic excitation is prefered for plastics and composite materials, this method appears to be less appropriate for the excitation of high mass concrete plates, at least at a significant vibration level.

The resulting vibrations of the plate are measured in a selected point by means of an accelerometer or by a laser vibrometer which measures the out-of-plane velocity of the vibrating surface.

The resonant frequencies of the T, S and B mode shapes may be determined by application of a simple peak-pick method or a curve fit algorithm on the Frequency Response Function (FRF). Recent efforts have been concentrated on time domain methods for modal damping extraction, based on curve fitting of the Impulse Response Function (IRF).

The whole identification procedure is directed by a menu-driven program that runs on a PC. The elastic moduli are first determined, because these values are needed to calculate the weighting functions for the contributions of the loss factors of the plate rigidities in equation (4).

5. Application to concrete-polymer composites

5.1. DESCRIPTION OF INVESTIGATED MATERIALS

The materials for which the testing results will be reported here, are the commercially available Cugla standard epoxy mortar and the Cuglaton polymer modified cementitious mortar.

The epoxy mortar is mostly used in repair applications and is produced from the mix of three basic components: the epoxy resin, the hardener and the aggregates (sand with max. grain size 0.6 mm). The mixing should be performed mechanically in order to obtain a homogeneous distribution of the sand within the epoxy resin. Compaction of the relatively thin plates cannot be obtained by means of a classical vibration table but may be performed by tapping with a trowel.

The polymer modified cementitious mortar is used in applications where high constraints are met regarding chemical resistance, water penetration, crack propagation... Good compaction results are obtained on a vibration table, although evacuation of air bubbles is not reached entirely.

5.2. DESCRIPTION OF TEST SPECIMENS

The dimensions of the test plates are carefully chosen in order to respect the Love-Kirchhoff displacement field conditions (the thickness is reduced up to 1/25 th of the largest dimension of the plate), to reduce the weight, but mainly to have resonant frequencies below 500 Hz, this in order to assure a sufficient signal to noise ratio with hammer excitation.

The plates, and accompanying prisms, are conserved at a temperature that is kept between 18°C and 20°C, and at a relative humidity between 66% and 70%.

5.3 EXPERIMENTAL RESULTS

Table 1 presents the results for the two types of investigated mortars, each material being characterized by the results for only one test specimen, which is however representative for all the other plates of the same material. The upper part of the table presents all the properties which are determined by measurement and which are used in the identification procedure to calculate the elastic moduli and their associated damping values, which are shown in the lower part of the table.

Table 1 : Material identification (19 days after production)

Material	epoxy mortar	polymer modified cementitious mortar
Plate identification Nr	DES 4	DPCC 4
Length (cm)	40.17	40.07
Width (cm)	39.97	40.00
Thickness (mm)	14.9	15.46
Mass (kg)	3.985	4.455
f_T (Hz)	114.9	174.7
f_S (Hz)	168.8	257.8
f_B (Hz)	194.9	289.7
η_T (%)	7.12	2.08
η_S (%)	6.86	2.08
η_B (%)	7.02	2.32
E (N/mm^2)	5 859	13 340
ν	0.21	0.17
G (N/mm^2)	2 424	5 676
η_E (%)	7.00	2.18
η_ν (%)	0.02	0.89
η_G (%)	7.00	2.05

Some comments may be formulated.

- The expressions (3) show that a small error on the measurement of the thickness t has important consequences as to the accuracy of the calculated engineering properties. Moreover, a constant thickness is rather difficult to obtain with a horizontal mould and the difficult spread out of the mortar. The open mould technique also leads, in the case of the epoxy mortar, to a top skin which is characterized by a high rugosity, with sand grains that are not entirely embedded into the mortar mass. A polishing operation should help in avoiding this problem. In order to produce plates with two flat sides and constant thickness, a vertical mould concept should yet be studied.

- Working with a coarse material necessitates the application of the identification method on several test specimens, in order to get mean values. The E-Moduli in table 1 are characterized by a standard variation of 1000 MPa on the mean value; damping values show to be less influenced by the relatively important variation on stiffnesses.
- The evolution of strength and stiffness (E and ν) has also been verified by means of classical static tests on prisms. The most important observation is that the E-modulus measured in quasi-static conditions is about 2000 MPa lower in value than the dynamically measured one, which is due to the viscoelastic behaviour that also determines the quasi-static properties.
- The viscoelastic effect is even more visible in the damping values, where the epoxy matrix is responsible for an increase of the loss factors; earlier studies on pure cementitious mortars showed much lower damping values ($\eta < 1\%$). It should be noted that, as in the case of fibre reinforced plastics [3], material damping as well as the elastic moduli, prove to be strongly affected by the degree of polymerisation of the resin matrix, the presence of water in the pores, and the ambiant temperature. It should also be noted that, within the pratical domain of confidence of the calculated results, the loss factors associated with the E and G moduli are the same, which may support the assumption of the Poisson-coefficient to be real valued.

6. Prediction of damping of reinforced microconcrete with short steel fibres

Short steel fibres (with length = 1cm) have been added to the microconcrete mix (up to 5% in mass), in order to produce thin square plates (50cm x 50cm x 2cm). Investigation of hardened plates showed up that all the fibres are randomly oriented, but always in a plane parallel to the midplane of the plate; this is due to the vibration of the mix by means of a horizontal shock-table. The fibre reinforced microconcrete may thus be considered as a transverse isotropic material.

The first step in the reasoning consists of the application of the above mentioned numerical model to determine the contribution of the stress-strain cycles σ_x - ε_x, σ_y - ε_y an τ_{xy} - γ_{xy} to the total elastic strain energy that is stored into a plate during vibration, for each mode shape seperately; the axes x and y are parallel to the edges of the plate. The first three modes are again the T, S and B modes. It is noted that the T mode shape is dominated by shear (94% τ_{xy} -γ_{xy}, 3% σ_x - ε_x and 3% σ_y - ε_y) while the S and B mode shapes are dominated by the normal stress-strain cycles (50% σ_x - ε_x and 50% σ_y - ε_y).

These figures take into account the Poisson-effect, which causes energy absorption or release related to a combination of stress with strain that is induced by stress in the perpendicular direction.

The second step in the reasoning consists of the evaluation of the loss factors associated with the micromechanic RVE-models of short fibre composites: η_1, η_2 and η_{12}. Several models may be proposed, all based on the well known COX shear lag model [4]. For η_1, one may write:

$$\eta_1 = \eta_f \frac{W_f}{W} + \eta_m \frac{W_m}{m} \tag{5}$$

In that way, the problem of η_1 is reduced to the determination of the elastic strain energy partition between fibres and matrix, taking into account that the fibres are loaded in pure tension while the matrix material is loaded in tension and shear: $W_m = W_{m\tau\text{-}\gamma} + W_{m\sigma\text{-}\varepsilon}$. The elaboration of the strain energy fractions on the basis of COX's stress equations leads to complex equations [5], in which the analytical expressions of the hyperbolic stress evolutions in the shear lag model, play an important role, as well as the assumptions regarding the fibre packing model. Analogous expressions may be developed for η_2 and η_{12}, when the RVE is loaded in tension perpendicular to the fibres and in shear. However, it should be emphasized that in view of the dominance of the steel fibres rigidity with respect to the matrix rigidity, the loss factor η_1 is essentially fibre dominated, while on the other hand, η_2 and η_{12} are matrix dominated.

The third step in the reasoning consists of defining the link between the strain energy distribution, determined in step 1, and the loss factors η_1, η_2 and η_{12} determined in step 2. This link is essentially determined by the introduction of the orientation of the fibres, and leads to the following partition of the modal loss factor:

$$\eta = \eta_{\sigma_x\text{-}\varepsilon_x} \frac{W_{\sigma_x\text{-}\varepsilon_x}}{W} + \eta_{\sigma_y\text{-}\varepsilon_y} \frac{W_{\sigma_y\text{-}\varepsilon_y}}{W} + \eta_{\tau_{xy}\text{-}\gamma_{xy}} \frac{W_{\tau_{xy}\text{-}\gamma_{xy}}}{W} \tag{6}$$

with:

$$\eta_{\sigma_x\text{-}\varepsilon_x} = \left(\eta_1 \frac{Sw_1}{W} + \eta_2 \frac{Sw_2}{W} + \eta_{12} \frac{Sw_{12}}{W} \right)_{\sigma_x\text{-}\varepsilon_x} \tag{7}$$

and analogous expressions for the other stress-strain combinations. η_1, η_2, η_{12} are the loss factors associated with the micromechanical models of a RVE with short fibres, subjected to longitudinal, transverse and shear loading (which have been discussed in step 2). $S_{Wij/W}$ designates the total fraction of energy which is stored into the three respective micromechanical models for all the different fibres, taking into account a randomly distributed fibre orientation. The symbol S is used because this fraction of energy is obtained by calculation of the surface under the curves W_{ij}/W versus orientation angle, for which the continuous fibre model is very well suited.

Figure 2 represents such a set of curves for the normal stress-strain combination σ_x - ε_x [3]. The value of S is determined by elastic material properties of fibres and matrix, by the fibre volume fraction (v_f) and by the distribution of fibre orientation.

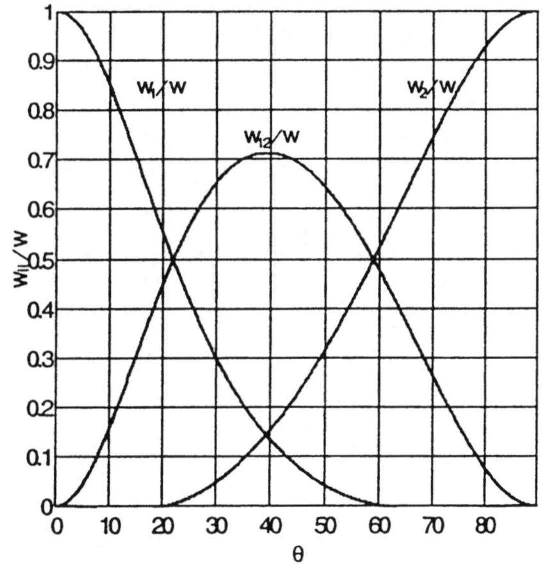

Figure 2 : Elastic strain energy fractions W_1/W, W_2/W and W_{12}/W versus orientation angle θ.

In order to get an order of magnitude of the S values, the case of ordinary microconcrete (from step 2) leads to the following partition of η:

$$\begin{aligned}\eta = &\left(0{,}34\eta_1+0{,}35\eta_2+0{,}31\eta_{12}\right)\frac{W_{\sigma_x-\varepsilon_x}}{W}\\ &+\left(0{,}34\eta_1+0{,}35\eta_2+0{,}31\eta_{12}\right)\frac{W_{\sigma_y-\varepsilon_y}}{W}\\ &+\left(0{,}23\eta_1+0{,}26\eta_2+0{,}51\eta_{12}\right)\frac{W_{\tau_{xy}-\gamma_{xy}}}{W}\end{aligned} \tag{8}$$

This expression proves to be of practical use in determining the modal loss factors, but also in evaluating the influence of fibre reinforcement on damping. Indeed, recalculation of the S values for a microconcrete with higher v_f leads to a general decrease of the contribution of η_1, and an increase of the contributions of η_2 and η_{12}, thus leading to a higher damping capacity. This last result confirms the experimental observation of increasing damping with increasing v_f. It should be noted that this conclusion is totally different from the one for continuous fibre reinforcements, where damping decreases with increasing v_f.

7 Conclusions

It is shown that the application of a mixed numerical-experimental technique leads to a full characterization of polymer modified mortars. Moreover, the identification method appears to be sensitive as to the effect of the hardening parameters as well as to the presence of a polymer matrix component, which results in a pronounced softening effect on the elastic moduli and an increased damping behaviour. The presented identification method offers the advantage of a quick and simple measurement method, that permits to study the hardening parameters in a much more versatile way than with quasi-static test methods.

The paper also presents a theoretical model for the evaluation of material damping of fibre reinforced microconcrete. The model is based on the energy balance approach which reduces the evaluation of damping to a problem of strain energy partition calculation, and the decomposition of the global composite damping phenomenon into three basic "micromechanical" damping mechanisms. In the case of short fibres, two additional problems are encountered: the somewhat precarious elaboration of the micromechanical models and the modelling of the random orientation effect. The last problem is solved by studying the energy balance in function of fibre orientation, while taking into account a continuous fibre reinforcement.

8 References

1. Cardon, A.H., Sol, H., De Wilde, W.P., Vantomme, J., Hongxing, H., De Visscher, J., *Mixed numerical-experimental techniques : recent developments and potentials for the future of experimental mechanics*, 10 th. Int. Conference on Experimental Mechanics, Lisboa, Portugal, 1994.
2. Sol, H., *Identification of anisotropic plate rigidities using free vibration data*, PhD thesis, Free University of Brussels VUB, 1986.
3. Vantomme, J., *A parametric study of material damping in glass fibre reinforced thermosetting plastics*, PhD thesis, Free University of Brussels VUB, 1992.
4. Bentur, A., Mindess, S., *Fibre reinforced cementitious composites*, Elsevier Applied Science, 1990.
5. Habiyaremye, F., *Etude de l'amortissement du béton armé de fibres et modelisation de l'erreur par l'approche de Bayes*, Engineering thesis, Royal Military Academy Belgium, 1993.

EXPERIMENTS AND SIMULATIONS OF INTERFACE FRACTURE IN CONCRETE

ADRI VERVUURT[1] AND JAN G.M. VAN MIER[2]
1. *TNO bouw*
PO Box 49
2600 AA Delft
2. *Delft University of Technology*
PO Box 5048
2600 GA Delft

ABSTRACT

At the aggregate level of concrete the fracture processes in the material are dominated by failure of the individual composites. A numerical lattice model has been applied which uses this knowledge for simulating fracture of concrete. In the model concrete is treated as a three phase material consisting of aggregates, matrix and the interfacial transition zone between the matrix and the aggregates. In order to determine the model's fracture properties with respect to the interfacial transition zone, experiments and simulations have been carried out simultaneously. Two dimensional tensile splitting tests have been performed in which a cylindrical aggregate particle was embedded in a matrix of cement paste or mortar. In the simulations a variation was applied to the properties of the beams in the lattice model. It was concluded that the fracture behaviour is dominated by the strength of the beams in the interfacial transition zone, whereas the beam stiffness seems to have a minor effect on the fracture mechanism.

1. Introduction

For heterogeneous materials like concrete it is widely accepted that the interfacial transition zone is the most critical region for crack initiation. When studying the interfacial transition zone in concrete it is essential that an unambiguous definition of this zone is given. In this respect it is noted that in this paper the mechanical properties of the interfacial transition zone between the aggregate particles and matrix (consisting of the hardened cement paste and the finest sand particles) are studied. From this definition it may be obvious that the research focuses on the particle level, also called meso level. Both at the meso level and the micro level interfaces are formed by a combination of physical, chemical and mechanical bonding. At the meso level

H. Sol and C.W.J. Oomens (eds.), Material Identification Using Mixed Numerical Experimental Methods, 153-162.

mechanical bonding seems to be dominating, whereas on the micro level physical and chemical bond is more pronounced (Zhang & Gjørv, 1990). At the macro level, on the other hand, concrete can be treated as a continuum material and no interface can be distinguished within the material. To describe the material behaviour at the latter level properly, a non linear constitutive material behaviour has to be assumed. It will be shown in this paper that at the meso level brittleness of the individual composites still leads to a non linear material behaviour. A difficulty in this approach is, however, that the properties of the modelled phases have to be known. For the aggregate and matrix phase this behaviour can be determined rather straightforward, but for the interfacial transition this is not that unequivocal. An often used method to study the mechanical properties of the interfacial transition zone is to isolate the aggregates by adopting only a few particles, which are positioned carefully at a prescribed position in a specimen (Lee et al., 1992), or by casting cement paste to a cored aggregate (e.g. Alexander (1959), Wang & Maji, (1994)). From the approaches followed by these researchers it appeared that there is still some disagreement in the exact role of the interfacial transition zone on the fracture behaviour of concrete. Moreover, it appeared that the conditions of preparing, curing and testing the specimens is of major importance when evaluating the results. In the experiments and simulations presented in this paper careful attention was paid to consistency in preparing the specimens and carrying out the tests. In the simulations it was tried to model the tests as accurate as possible. The fracture properties required in the numerical model have been determined by carefully observing the fracture processes in the experiments with an optical microscope.

2. Experiments

2.1 GEOMETRY AND MATERIALS

Experiments have been carried out on specially prepared specimens in which a cylindrical aggregate with a diameter of 20 mm is embedded in a cement paste or mortar (Figure 1a). For the matrix three different hardened cement paste mixtures were studied along with four types of mortar. Because the main goal was to learn more about the properties of the interfacial transition zone and because this zone is highly affected by the type of aggregate, two extreme types of aggregate materials were studied. In order to mimic a concrete with gravel aggregate particles, a granite cylinder was cored and embedded in the matrix. To create a dense interfacial transition zone, a porous aggregate cylinder was embedded in the matrix. For the latter case Bentheimer sandstone from Germany was adopted.

The specimens, with planar dimensions 125x170 mm^2 and a thickness of 15 mm, are loaded by a perfectly horizontal splitting load, as indicated in Figure 1a. For this purpose two loading platens are glued to the side and the top of the specimen. A 5 mm wide notch is sawn in the specimen with a depth of 30 mm, in order to fix the position of initial crack growth. A special set up was developed in which the specimens could be tested. The position of the aggregate with respect to the notch is also indicated in Figure 1a. The influence of the aggregate position on the specimen behaviour is reported elsewhere (Vervuurt, 1997). The tests were carried out under displacement control of the average Crack Mouth Opening Displacement (CMOD). The CMOD was measured across the notch (measuring length 15 mm) at the front and the rear side of the specimen.

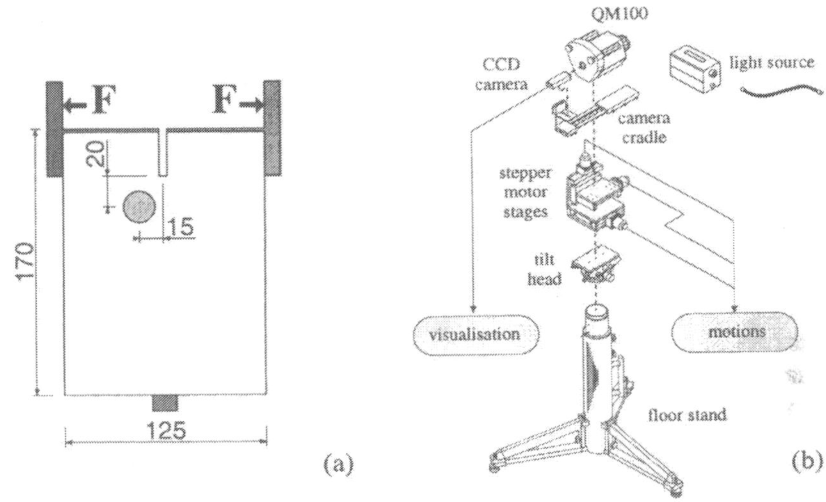

Figure 1 Specimen geometry (a) and a schematic view of the optical microscope (b).

In order to study the interfacial transition zone at a level comparable to the level of the numerical simulations, an optical microscope was used. This Remote Measuring System (Figure 1b) consists of a QUESTAR QM-100 microscope which is fastened to a camera cradle which can be moved in three orthogonal directions by means of stepper motors. The microscope is positioned in front of the specimens and images of crack growth near the interfacial transition zone are recorded during testing. Dedicated software was developed for controlling the movements of the microscope. Main goal for developing the new software was computerising the procedure for scanning the crack, and simultaneously storing the crack path and the corresponding images at the computer. After testing mosaics are created consisting of several images in order to receive a global overview of the cracks which have developed in the specimen. The Remote Measurement System, the developed software and the image processing techniques used, are elucidated in detail in Vervuurt (1997).

2.2 RESULTS

To clarify the most significant results of the experiments, two characteristic examples are discussed. The behaviour of the specimens is presented by means of the load-CMOD curve, the final crack pattern, and a crack detail near the interfacial transition zone, as recorded by the optical microscope. First the results of a granite aggregate embedded in a mortar matrix will be discussed and secondly the results are shown of a sandstone aggregate embedded in a similar matrix.

When a granite aggregate particle is embedded in a matrix, the situation seems representative for a concrete containing river gravel. In both cases the aggregate acts strong compared to the matrix, and the bond between the aggregate and the matrix is dominated by a weak interaction. For concrete it is therefore well accepted that cracking is initiated in the interfacial transition zone, and followed by failure of the matrix. In the experiments presented in this paper, however, a somewhat different behaviour is observed. Due to the selected geometry and the rather isolated location of the aggregate, cracking starts at the notch. In the load deformation curve (Figure 2a) this moment of crack initiation is indicated by the steep drop of the load in the load-CMOD curve. The reason that the specimen containing a granite aggregate a shows lower peak than the specimen of plain matrix material (as shown by the dotted line in the figure) is that stress concentrations appear due to the presence of the aggregate. It was observed with the microscope that, after cracking had started, the crack arrests at the aggregate particle immediately after the first drop of the load. Thereafter the loads increases again and microcracks develop in the interfacial transition zone. As soon as the load-CMOD curve has reached the envelope of the specimen containing no cylindrical aggregate, a second peak is observed in the load-CMOD curve, indicating crack deflection (Figure 1c). Almost at the same time the crack kinks out of the interfacial transition zone, after which cracking proceeds gradually to the edge of the specimen. The final crack pattern corresponding to this behaviour is given in Figure 2b.

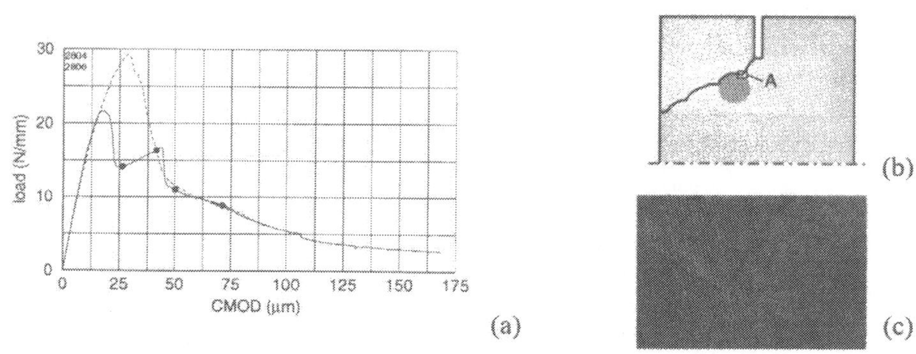

Figure 2 Load-CMOD curve (a) and final crack pattern (b) for a specimen containing a granite aggregate particle. The dotted line refers to a specimen containing no cylindrical aggregate. In (c) crack deflection in area A is shown.

Aggregates adopted in light weight concrete generally have a higher porosity and a lower strength than the granite as presented above. In the experiments with a single cylindrical aggregate a light weight aggregate particle was modelled by means of a sandstone aggregate particle. The results of such a test are shown in Figure 3. Contrary to the tests with granite, the crack is more likely to penetrate the interfacial transition zone (Figure 3b and Figure 3c). Moreover, a much higher second peak is observed in the load-CMOD curve (see Figure 3a). This can be explained by the fact that the weak aggregate acts as a circular *pore*. As soon as the crack has penetrated the aggregate particle (which occurs before the second peak is reached), a strain release is noticed due to the low stiffness of the aggregate. Thereafter stress concentrations appear at the position where the crack will leave the aggregate particle. Because of the circular shape of the matrix at that point (due to the aggregate geometry), a lower stress peak will be formed when compared to a perfectly sharp crack tip, resulting in a higher (second) peak load.

It is mentioned that the failure mechanism as explained above was not consistent in the tests but strongly depended on the microstructure of the interfacial transition zone. Due to the selected geometry and the procedure of casting and compacting the specimens, the matrix did not always flow properly around the aggregate particle, resulting in possibility of air inclusions in the interfacial region. It may be obvious that such air inclusions stimulate crack deflection. Unfortunately, in most of the cases this could be observed only after failure of the specimen.

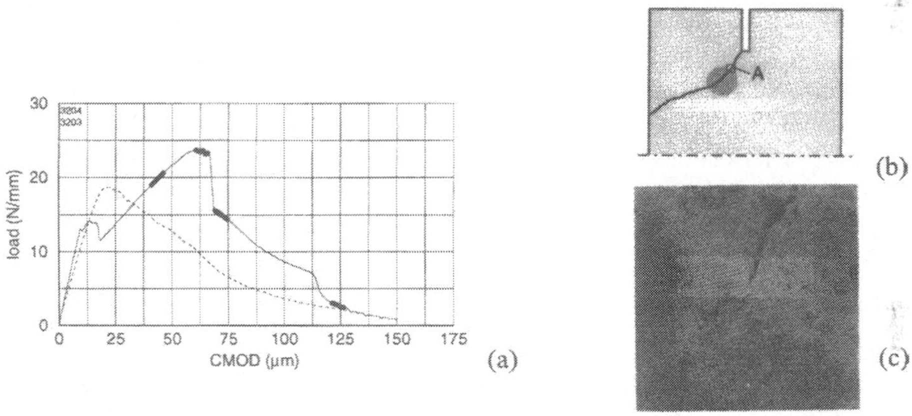

Figure 3 Load-CMOD curve (a) and final crack pattern (b) for a specimen containing a sandstone aggregate particle. The dotted line refers to a specimen containing no cylindrical aggregate. In (c) crack penetration in area A is shown.

3. Simulations

3.1 OUTLINE OF THE MODEL

A numerical lattice model has been adopted for simulating the experiments containing a single aggregate particle, as described in the previous section. In the model, which was originally proposed by Herrmann & Roux (1990) and adopted for simulating fracture in concrete by Schlangen & Van Mier (1992), the material is schematised as a network of brittle breaking beam elements. To simulate failure of the material, a beam is instantaneously removed from the mesh as soon as the strength of that particular element is exceeded. For modelling heterogeneous materials like concrete, disorder is implemented in the mesh. Several methods can be distinguished, depending on the type of mesh. When the network consists of a regular configuration of the beams, a variation of the strength and stiffness has to be assigned to the beams in order to model heterogeneity. For concrete it seems logical to use the material structure for assigning these varying material properties to the beams. Depending on the level of observation with respect to the size of the beams, the material is treated as a three phase material, consisting of the aggregates, the matrix (hardened cement paste and the fine sand particles), and the interfacial transition zone between the aggregates and the matrix. In each phase, the material is treated homogeneously, and the beams are therefore assigned with the same strength and stiffness. It will be shown that the heterogeneity of the material is characterised by the strength and stiffness ratios of the three material phases. For the strength (f_t) these ratios are $f_{t;M}/f_{t;A}$ and $f_{t;M}/f_{t;B}$, whereas for the stiffness the ratios are $E_{t;M}/E_{t;A}$ and $E_{t;M}/E_{t;B}$., where the subscripts A, M and B stand for aggregate, matrix and bond between matrix and aggregate respectively. Next to varying the material properties of the beams, heterogeneity of the material can be mimicked by applying a geometric disorder to the mesh (i.e. by altering the beam length). In the random lattice adopted in this paper (Moukarzel & Herrmann, 1992) a regular square grid with cell size s is generated, after which in each cell of the grid a node is selected at random. Subsequently the (triangular) lattice configuration is determined by connecting always the three nodes which are closest to each other. The procedure for generating the lattice is illustrated in Figure 4a.

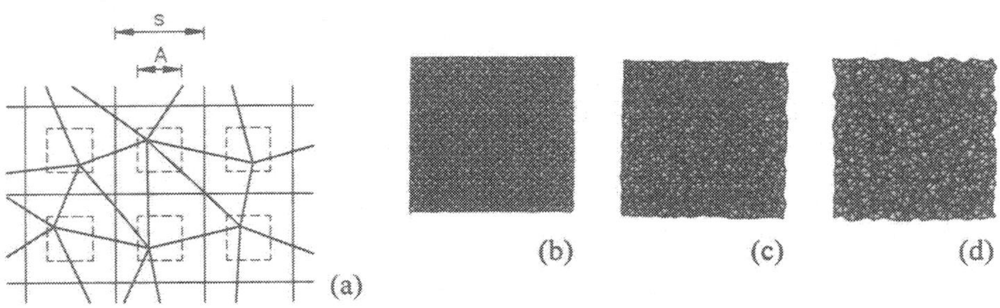

Figure 4 Generation of a random lattice (a) and meshes with a randomness of A=0.2·s (b), A=0.5·s (c), and A=0.8·s (d).

The randomness of the mesh is controlled by the parameter A, which limits the area in a cell where the node is selected. It may be obvious that for decreasing A, the randomness of the mesh decreases as well (as shown in the examples of Figure 4b-d), and that in the limit case of A=0, a regular square lattice is obtained with crossing diagonals. The random lattice adopted in this paper is assigned with a cell size of s=1 mm and a randomness of A=0.75·s. In spite of the fact that a random lattice with constant material properties for the beams behaves heterogeneously by nature, an aggregate overlay can be used to assign additional heterogeneity. In practise this means that each individual phase (i.e. aggregates, matrix and bond) is assumed to behave heterogeneously as well.

The geometric and material properties of the beams are the height *h*, the thickness *t*, the Poisson's ratio ν, and the Young modulus *E*, and depend only on the elastic behaviour of the lattice. The strength (f_t) of the beams, on the other hand, is directly related to the material's fracture behaviour. The determination of these properties can proceed in a rather straightforward manner as far as the beams in the matrix and the aggregate phase are concerned. The procedures for determining the material and geometric properties are explained in detail in Vervuurt (1997), and are not further discussed in this paper. It may be obvious, however, that the determination of the interfacial properties is not as straightforward as for the matrix and the aggregates, because the interfacial transition zone can not be isolated from the material. In the following paragraph the simulations are discussed in which the results of the (earlier mentioned) experiments are used for calibrating the interface properties in the lattice model.

3.2 STUDIED VARIABLES

In the presented simulations, the influence of varying the strength and stiffness of the beams in the interfacial transition zone between matrix and aggregates, on the fracture behaviour of a specimen, is studied. For this purpose the same geometry has been adopted as for the experiments presented earlier in this paper (i.e. Figure 1). A series of over 60 simulations was performed. In this paper, however, only the most significant results are presented. For a discussion on the complete series of simulations the reader is referred to Vervuurt (1997). Globally the simulations can be divided into two groups, namely those related to the experiments containing a granite aggregate particle and those with a sandstone aggregate particle. For each series three ratios $f_{t;M}/f_{t;B}$ and four ratios of $E_{t;M}/E_{t;B}$ were investigated. Next to varying the interfacial transition zone, a few additional parameters were studied, such as the mesh dependency and the position of the (big) aggregate particle. These results, however, are not discussed in this paper, but reported in Vervuurt (1997).

3.3 RESULTS

As mentioned above, only the most significant results are elucidated in this paper. It appeared that the most realistic results were obtained for granite with the following strength and stiffness ratios for interfacial transition zone: $f_{t;M}/f_{t;B}=4$ and $E_{t;M}/E_{t;B}=4$. The results of this simulation are given in Figure 5.Moreover, it will be shown that, for simulating the experiments containing a sandstone aggregate particle, it seems essential that the interfacial strength is larger than the strength of the matrix (i.e. $f_{t;B}>ft_{;M}$). Actually the results which fitted the experimental observations with respect to a sandstone aggregate particle best, were: $f_{t;M}/f_{t;B}=0.75$ and $E_{t;M}/E_{t;B}=1.5$. These results are shown in the Figure 6 and Figure 7.

The results for a specimen containing a granite aggregate particle are summarised in Figure 5 by means of the load-CMOD curve (Figure 5a) and the crack patterns during three stages in the fracture process (Figure 5b-d). To highlight the micro cracks which develop in the interfacial transition zone, only the beams which have been removed from the mesh are plotted. It appeared that cracking is initiated at the notch, after which crack deflection at the interfacial zone is observed. When about halve the aggregate's circumference is damaged, cracking proceeds again from the notch (Figure 5b). As soon as the two cracks coalesce, and the crack proceeds growing to the side of the specimen, the second peak in the load-CMOD curve is reached (Figure 5c). Thereafter, a gradual softening branch is observed. The behaviour is globally the same as observed in the experiments, except that in the experiments the crack seemed to grow until the aggregate's circumference before debonding was observed. Nevertheless, when a higher interfacial strength for the interface beams is selected, the second peak vanishes from the load-CMOD curve. It seems that interface cracking in the simulations is essential for obtaining the second peak in the load-CMOD response. Moreover it appeared that the influence of the stiffness of the beams in the interfacial transition zone can be neglected since the strength of the interface beams dominates the fracture process (Vervuurt, 1997).

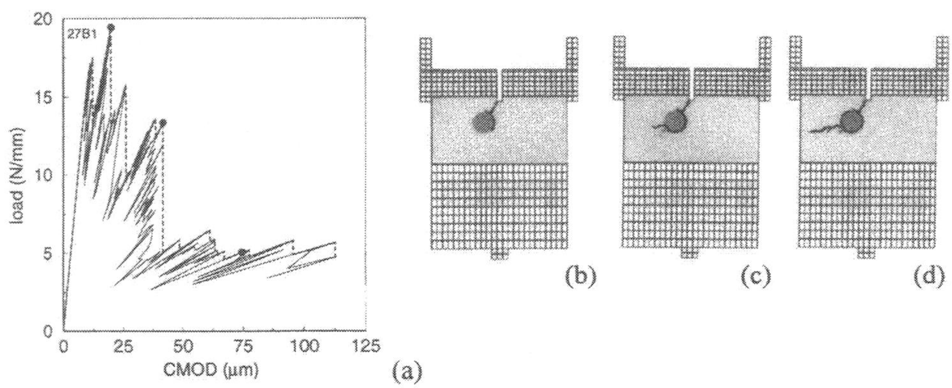

Figure 5 Load-CMOD curve (a) and crack history (b-d) for a simulation containing a granite aggregate particle. The black dots in the load-CMOD curve refer to the plotted stages of cracking.

For simulating the tests containing a single sandstone aggregate particle a considerably lower aggregate stiffness and a higher interfacial strength was adopted. It was found again that the strength of the beams in the interfacial transition zone has the largest influence on the final crack pattern and the load-CMOD curve, and that the ratio $f_{t;M}/f_{t;B}$ should always be chosen smaller than 1. This is in Figure 6 where the results are given for two extreme cases with varying interfacial strength (i.e. $f_{t;M}/f_{t;B}$=0.75 and $f_{t;M}/f_{t;B}$=1.25 respectively). The remaining ratios (with respect to the strength and stiffness) were kept constant for the two simulations. The final crack patterns in Figure 6b and 6c were retrieved at the end of the simulations. It can be seen clearly from the load-CMOD curve that the pronounced second peak as observed in the experiments, is formed only when $f_{t;M}/f_{t;B}$=0.75. Moreover, it appeared that the second peak was always accompanied by crack growth through the aggregate particle.

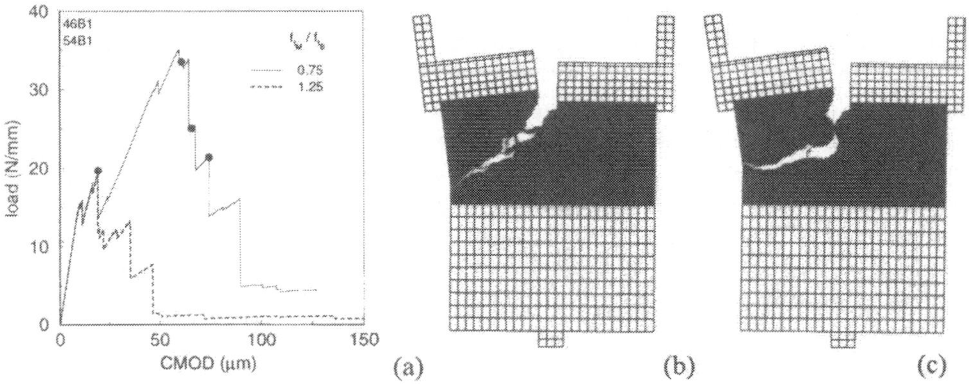

Figure 6 Load-CMOD curve (a) and final crack patterns for simulations with $f_{t;M}/f_{t;B}$=0.75 (b) and $f_{t;M}/f_{t;B}$=1.25 (c). The black dots refer to the stage of cracking in Figure 7.

In Figure 7 the history of cracking is given for the simulation of Figure 6 with $f_{t;M}/f_{t;B}$=0.75 (see Figure 6) by means of four crack patterns. It can be seen that, at the time the first peak is reached in the load-CMOD diagram (Figure 6a), the crack has grown until the aggregate's circumference. Thereafter a small drop of the load is observed. Ahead of the second peak, isolated micro cracks are formed around and perpendicular to the aggregate particle. After the second peak has been reached (Figure 7b-d) crack growth proceeds very quickly to the side of the specimen, which is consistent to the experimental observations.

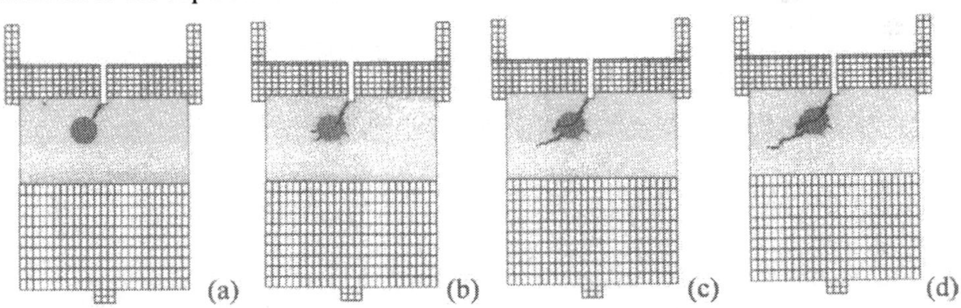

Figure 7 Crack history for a simulation with $f_{t;M}/f_{t;B}$=0.75, containing crack patterns with 25 (a), 75 (b), 150 (c), and 225 (d) removed beam elements. These stages correspond to the black dots in Figure 6.

4. Conclusions

A combined numerical and experimental approach is adopted for studying the mechanical properties of the interfacial transition zone between matrix and aggregate in heterogeneous materials like concrete. Splitting tensile experiments were performed in which a perfectly horizontal load was applied to a specimen containing a single, cylindrical aggregate particle, which was located at a prescribed position in the specimen. An optical microscope was adopted for monitoring crack growth in the interfacial region. In both the experiments and the simulations two types of aggregate with extreme interfacial properties were tested. A porous sandstone was used to imitate the behaviour of a light weight aggregate material whereas granite was used as a substitute for a gravel composite. From the experiments it appeared that the interfacial zone for granite acts very weak, which is reflected by crack deflection at the interfacial transition zone. For the sandstone aggregate particle the crack is more likely to penetrate the interface. In the simulations with the lattice model this behaviour is dominated by the strength of the beams in the interfacial transition zone. The beam stiffness seems to have a minor effect on the fracture mechanism.

5. Ackowledgements

The authors greatly acknowledge Mr. A.S. Elgersma for his expertise in building part of the experimental set up and carrying out the experiments.

6. References

Alexander, K.M. (1959) Strength of the Cement-Aggregate Bond. *Journal of the American Concrete Institute*, **56(11)**, 377-390.

Lee, K.M., Buyukozturk, O. and Oumera, A. (1992) Fracture Behaviour of Mortar-Aggregate Interfaces in Concrete. *Journal of Engineering Mechanics (ASCE)*, **118(10)**, 2031-2047.

Moukarzel, C. and Hermann, H.J. (1992) *A Vectorizable Random Lattice.* Technical Report HLRZ 1/91 (preprint), KFA Jülich.

Herrmann, H.J. and Roux, S. (1990). *Statistical Models for the Fracture of Disordered Materials.* Elseviers Science Publishers B.V.

Schlangen, E. and Van Mier, J.G.M. (1992). Experimental and Numerical Analysis of Micromechanisms of Fracture of Cement Based Composites. *Cement and Concrete Composites*, **14**, 105-118.

Vervuurt, A.H.J.M. (1997) *Interface Fracture in Concrete.* PhD thesis, Delft University of Technology.

Wang, J. and Maji, A.K. (1994) Experimental Studies and Modelling of the Concrete/Rock Interface. *ACI Special Publication on Interface Fracture and Bond*, **SP-156**, 45-68.

Zhang, M.H. and Gjørv, O.E. 1990. Microstructure of the Interfacial Transition Zone Between Light Weight Aggregate and Cement Paste. *Cement and Concrete Research*, **20**, 610-618.

INVERSE ANALYSIS OF A TUNNEL EXCAVATION PROBLEM FROM DISPLACEMENT AND PORE WATER PRESSURE MEASUREMENTS

A. LEDESMA, A. GENS
Technical University of Catalunya (UPC)
Geotech. Engng. Dept., ETS Ingenieros de Caminos
Jordi Girona, 1-3. 08034 Barcelona, SPAIN

In this paper a methodology to identify parameters of elastoplastic geomechanical models from displacement and pore water pressure measurements is presented. The methodology is formulated within a probabilistic framework, and the weight of each source of information on the identification process is obtained in a consistent manner. The formulation is applied to the backanalysis of a laboratory experiment, involving the excavation of a tunnel in clay.

1. Introduction

Model parameters involved in Geomechanical problems have been determined commonly by means of laboratory tests or "in situ" tests. Backanalyses of real problems have also been performed, but usually in an "ad hoc" manner for each particular case. However, recent development on numerical techniques to solve direct problems (i.e. finite element method) coupled with optimization methods, have permitted the formulation of new procedures to solve geomechanical backanalysis in a more systemmatic manner. Some of these works refer to hydrogeological modelling (Carrera and Neuman, 1986), but there are also recent applications to soil/rock stress – strain problems (Honjo *et al*, 1994; Ledesma *et al*, 1996a).

Although there have been significant contributions in this field using a deterministic point of view, there are important advantages in using a probabilistic approach (Gioda and Sakurai, 1987). In this paper, a probabilistic framework based on the concept of maximum likelihood is proposed. The formulation requires finally the minimization of an objective function which depends on the differences between measured and computed values. The model must be defined in advance, and it is considered fixed in the analysis. The formulation has been successfully used in previous works

H. Sol and C.W.J. Oomens (eds.), Material Identification Using Mixed Numerical Experimental Methods, 163-172.

to estimate Young's moduli and initial stress conditions in Geomechanics (Gens *et al*, 1996), and to incorporate prior information on the parameters in the identification procedure (Ledesma *et al*, 1996b).

This paper presents the extension of that formulation to the case in which a nonlinear constitutive law is used to represent the stress – strain soil/rock behaviour. Also, the procedure copes with the problem of defining the weight of different sources of information in the final identification process. This is easy to incorporate in the formulation when a probabilistic framework is considered. In this case, the ratio of variances of each group of measurements is included as an additional parameter to be identified. Previous works have shown the utility of this approach in characterizing the variances of each measurement. (Carrera and Neuman, 1986; Ledesma *et al*, 1996b).

Finally, an example of application is presented. It refers to a laboratory experiment carried out at Cambridge University (Mair, 1979), involving the simulation of a tunnel failure in a centrifuge equipment where soil displacements and pore water pressures at different points in the soil mass were measured during the test, and are used now as main data for the inverse problem.

2. Probabilistic formulation

Assume that a deterministic model relates a certain set of variables, $\mathbf{x}$, with a set of parameters, $\mathbf{p}$, that is: $\mathbf{x} = \mathbf{M}(\mathbf{p})$, where $\mathbf{M}$ represents the model which is defined usually by means of the finite element method. Differences between measured, $\mathbf{x}^*$, and computed values using this expression are assigned to an error measurement. The model is assumed to be correct in this analysis. The maximum likelihood method (Edwards, 1972), defines the best parameters as those that maximize the likelihood of a hypothesis, L, defined as proportional to the probability of reproducing the actual measurements. That is, assuming a gaussian distribution:

$$L = kP(\mathbf{x}) = \frac{1}{\sqrt{(2\pi)^m |\mathbf{C_x}|}} \exp\left[-\frac{1}{2}(\mathbf{x}^* - \mathbf{x})^t (\mathbf{C_x})^{-1} (\mathbf{x}^* - \mathbf{x})\right] \quad (1)$$

where $\mathbf{x}^* - \mathbf{x}$ is the vector of differences between measured and computed variables, m is the total number of measurements, and $\mathbf{C_x}$ is the covariance matrix of error measurements. Maximizing L is equivalent to minimize $S = -2\ln L$:

$$S(\mathbf{p}) = (\mathbf{x}^* - \mathbf{x})^t (\mathbf{C_x})^{-1} (\mathbf{x}^* - \mathbf{x}) + \ln|\mathbf{C_x}| + m\ln(2\pi) - 2\ln k \quad (2)$$

where last two terms are constant. In the example presented in this paper, there are two different groups of measurements (displacements and pore water pressures). Errors can be considered independent in each group, and therefore covariance matrices are diagonal, although each group has its own variance σ^2. Thus equation (2) can be written as:

$$S(\mathbf{p}) = \sigma_x^{-2}(\mathbf{x}^* - \mathbf{x})^t(\mathbf{x}^* - \mathbf{x}) + \sigma_u^{-2}(\mathbf{u}^* - \mathbf{u})^t(\mathbf{u}^* - \mathbf{u}) + m_x \ln \sigma_x^2 + m_u \ln \sigma_u^2 \tag{3}$$

where now σ_x^2 and σ_u^2 refer to the displacements and pore water pressures variance respectively. Also, m_x and m_u correspond to the number of displacements and pore water pressures measured.

Minimization of (3) requires to perform additional hypothesis on the variances of measurements. This is difficult to know in Geomechanics, as the measurement process needs different devices to be installed in the soil mass. Therefore, these variances have been incorporated in the identification procedure as additional parameters.

It can be shown that the minimum of (3) depends on the parameters and on the value $\mu = \sigma_x^2/\sigma_u^2$, which can be computed in an uncoupled manner from the classical identification of the parameters (Ledesma, 1996b). Expression (3) becomes then:

$$S = m + m_x \ln\left(\frac{J'}{m}\right) + m_u \ln\left(\frac{J'}{\mu m}\right) \quad ; \quad J' = J_x + \mu J_u \tag{4}$$

where J_x and J_u are the minimum of (3) when only one kind of measurements (either displacements or pore pressures) are available, and $m = m_x + m_u$.

The minimum of function (3) can be obtained using any suitable algorithm. Here an extension of the Gauss – Newton algorithm due to Marquardt has been used (Tarantola, 1987). First a value of μ is chosen. For that value, the minimum is estimated iterating from any point in the space of parameters and using:

$$\mathbf{\Delta p} = \left[\mathbf{A^t}(\mathbf{C_x})^{-1}\mathbf{A} + \Lambda\mathbf{I}\right]^{-1}\mathbf{A^t}(\mathbf{C_x})^{-1}\mathbf{\Delta x} \tag{5}$$

where $\mathbf{A} = \partial\mathbf{x}/\partial\mathbf{p}$ is called sensitivity matrix and Λ is a scalar that is introduced to accelerate the convergence of the algorithm. If Λ increases, the value of $\mathbf{\Delta p}$ tends towards the gradient direction. Close to the minimum Λ decreases reaching zero. Therefore, for each value of μ, expression (4) can be computed, and then by direct search the general minimum of (4) is obtained.

The sensitivity matrix is computed by means of a finite difference scheme:

$$\frac{\partial \mathbf{x}}{\partial p} \approx \frac{\mathbf{x}(p+\Delta p)-\mathbf{x}(p)}{\Delta p} \tag{6}$$

When a linear stress – strain constitutive law is adopted, computation of **A** can be performed in an exact way, using the finite element aproximation (Ledesma *et al*, 1996a). However, when a nonlinear constitutive model is used, it is difficult to extend that procedure and the finite difference scheme has proven to be successfull and simple when there are not too many parameters involved in the problem.

3. Application to a tunnel excavation in a centrifuge

3.1. DESCRIPTION OF THE TEST

The methodology outlined above has been used to characterize the clay used by Mair (1979) in the simulation of a tunnel failure using a centrifuge equipment. Mair carried out a few tests in order to investigate the effect of different conditions (geometry, stresses, etc) on a tunnel excavated in clay. Displacements and pore water pressures generated during the test were measured in some points.

The material used was a kaolin that was mixed with water reaching a 160 % of water content. Then it was placed in the centrifuge box and a few load steps were applied, allowing for drainage in each case. After unloading, installation of displacement and pore water pressure transducers was performed. Then the tunnel was cut and a membrane was installed as a tunnel lining. An inner pressure σ_t was applied in order to recover the initial stress state.

The model reached a constant acceleration of 75g when the test was performed. These operations produced an overconsolidation of the clay, with almost a constant undrained shear strength profile about $25KN/m^2$. However, initial stress conditions were quite complex in order to get that constant value (Mair, 1979).

At a constant velocity of the centrifuge, the inner pressure on the tunnel membrane was $156KN/m^2$, which corresponds to the initial value of the vertical stress in the center of the tunnel. That pressure was reduced step by step until failure. For values $\sigma_t = 107$, 92 and 76 KN/m^2, surface and crown displacements were measured. Also pore water pressures were recorded during the whole test. Figure 1 shows the finite element mesh of the prototype used in this analysis (half of the mesh due to symmetry) and the position of the pore water pressure transducers.

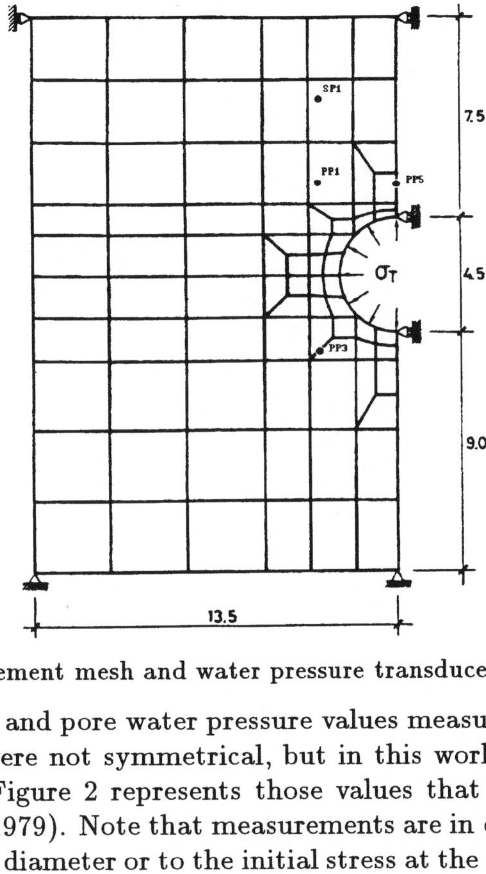

Figure 1. Finite element mesh and water pressure transducers. Dimensions in m.

Displacements and pore water pressure values measured in the test used in this analysis were not symmetrical, but in this work, an average value has been used. Figure 2 represents those values that correspond to test 2DP from Mair (1979). Note that measurements are in dimensionless form, referred to tunnel diameter or to the initial stress at the center of the tunnel ($156KN/m^2$). Water pressures on the tunnel crown have not been used in this analysis, because aparently the transducer did not work properly.

3.2. SOIL MODEL

The elastoplastic model used here is a modification of the Cam – clay model. A detailed description of the model adapted to the finite element method can be seen in Britto and Gunn (1987). Mair carried out different laboratory tests to determine the basic properties of the clay involved in the simulation using that model as reference. According to his tests, consistent values of the isotropic consolidation line for loading ($\lambda = 0.3$) and unloading ($\kappa = 0.05$) have been considered as known in this analysis. Also, the specific volume for a stress of $1KPa$ was assumed to be $v = 3.92$.

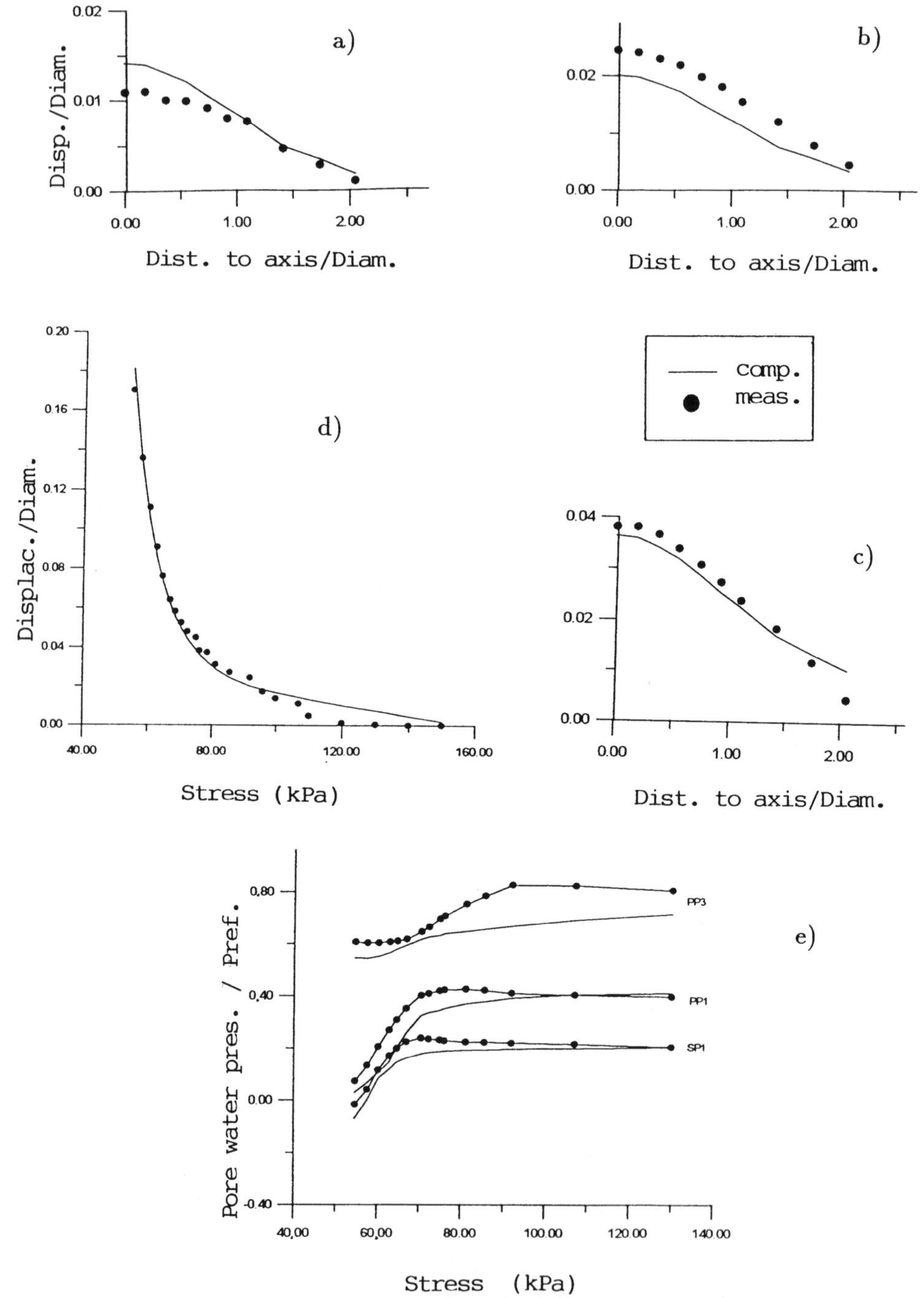

Figure 2. Computed and measured values. Surface settlements for (a) $\sigma_t = 107kPa$, (b) $\sigma_t = 92kPa$, (c) $\sigma_t = 76kPa$. (d) Displacements at crown for different σ_t values. (e) Pore water pressures for different σ_t values. $P_{ref} = 156kPa$.

In the Cam – clay model, failure conditions are obtained when the effective stress path reaches the critical state line and there are plastic strains. The slope of this line is defined as M in the mean effective stress – deviator stress plane. On the other hand, elastic strains are computed from bulk modulus, K, and shear modulus, G, that depend on mean effective stress. The ratio G/K (controlling elastic conditions) and the value of M (related to failure conditions) have been assumed as parameters to be identified.

3.3. IDENTIFICATION OF PARAMETERS

The information provided by the test consisted of 53 displacements and 45 pore water pressures. They correspond to different points and different stress σ_t. A finite element program (ICFEP) was used to reproduce the direct test. Cam – clay model and associated plasticity was used as constitutive law. Only the main features of the analysis are presented here. A detailed description of the identification procedure is presented in Rosas (1993) and Galopa (1997).

First, the inverse problem has been solved using only measured soil displacements as input data. As there are only two parameters to identify, it is possible to depict the path followed by the identification procedure (figure 3a). The values of the parameters identified correspond to $X = M/\sqrt{3} = 0.21$ and $G/K = 0.20$. In figure 2 displacements and water pressures computed with these parameters are also presented.

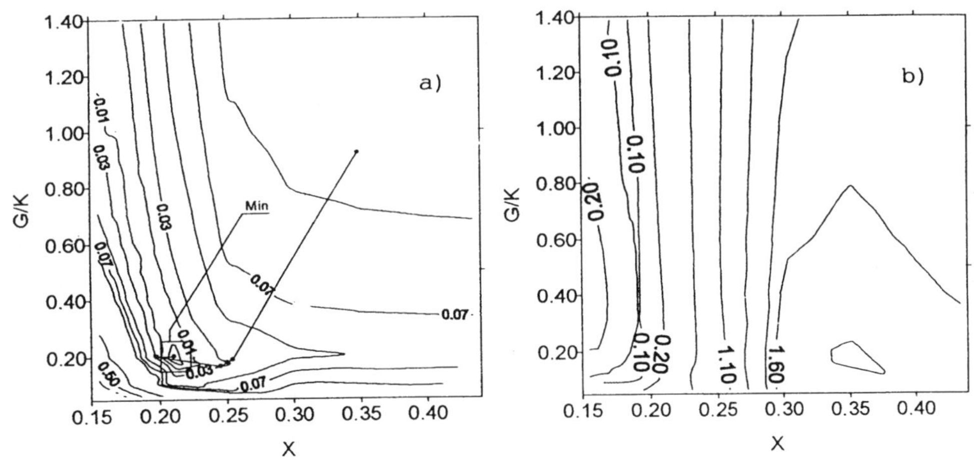

Figure 3. (a) Contours of objective function using displacements only. A path followed by the iteration procedure is indicated. (b) Contours of objective function using pore water pressures only.

When only measurements of pore water pressure are considered, the objective function to minimize is different, and it is shown in figure 3b. Note that in this case level curves are almost parallel to the G/K axis, which results in a difficult estimation of this parameter. In fact, that shape indicates that water pressure does not depend too much on the soil elastic parameters. This is important in this test, where the more important measurements are obtained close to failure (far from elastic conditions). In this case minimum is located close to the limits $X = 0.15$, $G/K = 0.15$.

If both set of measurements are used in the analysis, μ must be included as additional parameter, in order to estimate the relative weight of each set. Following the process above mentioned, a value of expression (4) is computed for each value of μ. Figure 4a presents the results obtained for different values, and indicates that the minimum corresponds to $\mu = 0.01$ (dimensionless value). Note that $\mu = \sigma_x^2/\sigma_u^2$, and therefore the process assigns a variance to the displacements which is 100 times smaller than the pore water pressure variance. However, values of J_x and μJ_u are similar, which indicates that water pressures give information to the analysis, although the final objective function, $J_x + 0.01J_u$, presented in figure 4b is not too different of function from figure 3a, corresponding to the case in which only displacements are considered. Therefore, the final parameters identified using all the information available are $X = 0.21$, $G/K = 0.20$ as in the first analysis.

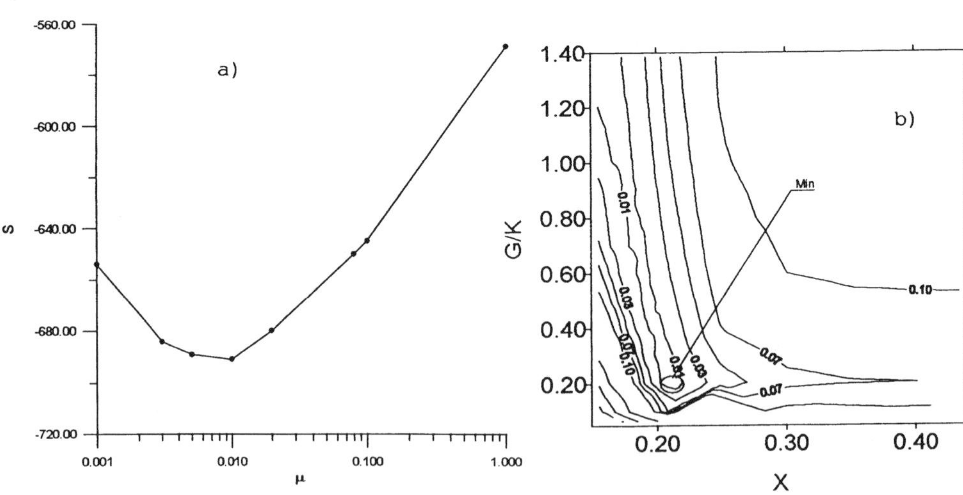

Figure 4. (a) Minimum of the objective function for each variance ratio μ. (b) Contours of objective function using displacements and pore water pressures.

The parameters obtained seem to be consistent with the information available. However, the value of X, which controls failure conditions, is smaller than the value expected for this kaolin, according to other laboratory results. Mair (1979) indicated that in this test, failure ocurred before expected and that can explain the smaller value identified.

On the other hand, the inverse analyses performed indicate that pore water pressure give information to the identification process, but they do not discriminate with respect to the analysis in wich only displacements are considered. That is, the identification is mainly controlled by displacements.

Nevertheless, this result can not be assumed as a general condition of this model. There are some factors that have influence on this result: First the value of the parameters, that change the shape of the objective function. Secondly the fact that in this test soil is overconsolidated, and stress paths when decreasing σ_t are mainly vertical in the mean stress – deviator stress plane. That is mean stress (total and effective) does not change significantly during elastic conditions, and therefore pore water pressure remains almost constant in this part of the test.

4. Conclusions

In this work a methodology to perform backanalysis in Geomechanics, using elastoplastic models, when two different measurement groups are involved, has been presented. The finite element method has been used to solve the direct problem, and Marquardt algorithm has been applied as minimization procedure.

The procedure has been defined in a probabilistic framework. This is useful when two different sources of information are available. In this case, the weight of each source (that is its variance) is difficult to know in advance, and to overcome this difficulty, the variance ratio of the measurements has been included as additional parameter to be identified. In this way a result consistent with the whole procedure has been obtained.

An application of the methodology to the identification of two parameters of a Cam – clay model (one related to failure conditions and another one linked to elastic conditions) from displacements and pore water pressures measured in a centrifuge simulation of a tunnel failure (Mair, 1979) has been presented. The inverse analysis shows that for this particular case, pore water pressures do not modify the parameters identified using only displacements. However this result is not general, and is due mainly to the stress paths followed in this problem.

Acknowledgements

The support given by "Departament de Política Territorial i Obres Públiques (Generalitat de Catalunya)" and by "Colegio de Ingenieros de Caminos, Catalunya" is gratefully acknowledged.

References

Britto,A.M. and Gunn,M.J. (1987) *Critical State Soil Mechanics via Finite Elements.* Ellis Horwood Ltd, Chichester, UK.

Carrera,J. and Neuman,S.P. (1986) Estimation of aquifer parameters under transient and steady state conditions. 1, Maximum likelihood method incorporating prior information. 2, Uniqueness, stability and solution algorithms. 3, Application to synthetic and field data. *Water Resources Res.*, 22, 199 – 242.

Edwards, A.W.F. (1972) *Likelihood*, Cambridge University Press, Cambridge, UK.

Galopa,J.M. (1997) *Análisis inverso de la excavación de un túnel a partir de medidas de desplazamientos y presiones de agua.* M.Eng. thesis. School of Civil Engng. UPC, Barcelona, Spain.

Gens,A., Ledesma,A. and Alonso, E.E. (1996) Estimation of parameters in Geotechnical backanalysis - II. Application to a tunnel excavation problem. *Computers and Geotechnics*, 18, 1, 29 – 46.

Gioda,G. and Sakurai, S. (1987) Back analysis procedures for the interpretation of field measurements in geomechanics. *Int. Journal for Numerical and Analytical Methods in Geomechanics*, 11, 555 – 583.

Honjo,Y., Wen-Tsung, L. and Guha, S. (1994) Inverse analysis of an embankment on soft clay by extended bayesian method. *Int. Journal for Numerical and Analytical Methods in Geomechanics*, 18, 709 – 734.

Ledesma,A., Gens,A. and Alonso, E.E. (1996a) Estimation of parameters in Geotechnical backanalysis - I. Maximum likelihood approach. *Computers and Geotechnics*, 18,1, 1 – 27.

Ledesma,A., Gens,A. and Alonso,E.E. (1996b) Parameter and variance estimation in Geotechnical backanalysis using prior information. *Int. Journal for Numerical and Analytical Methods in Geomechanics*, 20, 119 – 141.

Mair,R.J. (1979) *Centrifugal modelling of tunnel construction in soft clay*, PhD thesis, Cambridge University, UK.

Rosas,J.J. (1993) *Identificación de los parámetros de un modelo elastoplástico. Aplicación a la excavación de un túnel.* M.Eng. thesis. School of Civil Engng., UPC, Barcelona, Spain.

Tarantola, A. (1987) *Inverse problem theory.* Elsevier, Amsterdam.

CHARACTERIZATION OF THE IN-PLANE MECHANICAL BEHAVIOUR OF THE HUMAN SKIN IN VIVO

A Mixed Numerical-Experimental Approach Employing a Structural Skin Model

W.K.L. VAN DER VOORDEN
Eindhoven University of Technology
Postbus 513, 5600 MB Eindhoven the Netherlands

AND

L.F.A. DOUVEN
Personal Care Institute, Philips Research Laboratories
Prof. Holstlaan 4, 5656 AA Eindhoven, The Netherlands

Abstract. The subject of this study is the estimation of parameters in the constitutive model describing the mechanical behaviour of human skin in vivo. In the past, it was shown that indentation of skin can be modelled with a nonlinear, isotropic (hyperelastic) constitutive equation (Douven et al., 1994). However this model fails when it is applied for loads that are parallel to the surface of the skin. A more appropriate constitutive model for the skin is Lanir's Skin Model (Lanir, 1983). A mixed numerical experimental technique is applied to determine some of the parameters in Lanir's Skin Model.

1. Introduction

In section 2 the mechanical behaviour of skin is related to its structure. In section 3 the essential features of Lanir's Skin Model are described. The different elements of the strategy to identify parameters in this model from experimental data is given in section 4. In section 5 the results of parameter estimation from experimental data is described. Finally, some conclusions are drawn in section 6.

2. Skin: a complex material

From an anatomical point of view, skin is often defined as a fibrous connective tissue. From an engineers point of view it can be seen as a very complex composite material. The skin has a stratified structure where three main layers can be distinguished.

H. Sol and C.W.J. Oomens (eds.), Material Identification Using Mixed Numerical Experimental Methods, 173-182.

The top layer is called epidermis. This is a rather stiff but very thin layer. The second layer, the dermis, is a dense fibrous connective tissue layer and is generally an order of magnitude thicker than the epidermis. It consists of cells, groundsubstance and fibres. The fibres form a dense network and are surrounded by the groundsubstance. The third layer, the hypodermis, is a fatty layer of loose connective tissue.

The mechanical behaviour of skin depends primarily on the response of the constituents of the dermis (Lanir, 1983). Its main constituents are collagen and elastin fibres embedded in a matrix of amorphous ground substance. Collagen accounts for about 75% of the fat-free dry weight (Brown, 1972), (Manschot, 1985). Morphological observations (Finlay, 1969), (Ridge *et al.*, 1966) suggest that the collagen fibres are arranged in a three dimensional but primarily planar wavy array (Finlay, 1969). This waviness or undulation is considered as an important feature in skin mechanics. Upon stretch the collagen fibres become gradually straight, until they straighten completely. This process is accompanied by an increase of stiffness of the skin. The elastic modulus of collagen is approximately $1.10^8\ N{\cdot}m^{-2}$ (Savenije, 1982).

Elastin fibres are less abundant in skin than collagen and cover approximately 4% of the fat-free dry weight (Brown, 1972). The elastic modulus of elastin is approximately $1.10^6\ N \cdot m^{-2}$ (Savenije, 1982).

The ground substance is a complex matrix and can be characterized as a semi-fluid amorphous material. Its main constituents are glycoaminoglycans (GAG) and bind the present water in a molecular form. The ground substance is also thought to be partly responsible for the viscous part of the mechanical response of skin.

From the previous part it is clear that the mechanism of deformation of skin is a very complex one; it depends on the alterations in, and the interactions between its structural components. Fibres unfold, rotate and stretch and with that they squeeze the matrix that reacts through a buildup of pressure. Summarizing it can be stated that the overall response of skin to macroscopic deformation contains the following effects: nonlinearity, viscoelasticity, anisotropy and inhomogeneity.

3. Lanir's skin model

Lanir's Skin Model (Lanir, 1983) is a constitutive skin model employing a structural approach. This 3-D anisotropic nonlinear skin model is based on the assumption that the tissue's response is the sum of the responses of its constituents. Hence, if the constituents' structure, their mechanics and interactions are known, then the overall tissue response can be evaluated. In the following paragraph some definitions and assumptions of Lanir's Skin

Model are stated.

The mechanical behaviour of skin is assumed to be dominated by the dermis; the influence of the epidermis and hypodermis is neglected. In this model structure is defined in terms of the fibres' orientation. To each type of fibre, k, a (discrete) density distribution function, $R_k(\vec{r}_0)$, is associated, where $\vec{r}_0$ is a unit vector tangent to the fibre in reference configuration. This leads to a mathematical approach based on the following assumptions:

- Each fibre is thin and perfectly flexible. It has no compressive stiffness and it does not bear any load when it is undulated.
- If a fibre is stretched, it is subjected to a uniaxial strain which is the tensorial transformation of the overall strain in the fibre's direction.
- The fibres are elastic.
- Skin is incompressible.
- The effect of the matrix due to deformation is that of a hydrostatic pressure.
- Upon stretching the fraction of fibres that are straightened and stretched rises, providing nonlinear behaviour.

The stress-strain relation is described with:

$$\boldsymbol{\sigma} = \boldsymbol{\tau} - p\boldsymbol{I}, \tag{1}$$

with $\boldsymbol{\sigma}$ the Cauchy stress tensor, $\boldsymbol{\tau}$ the so-called extra stress tensor, p a hydrostatic pressure and $\boldsymbol{I}$ the second order unit tensor. The mechanical behaviour of the fibres is accounted for in $\boldsymbol{\tau}$, the contribution of the groundsubstance is accounted for in p. The stress-strain relation is then defined by (Feron, 1993):

$$\boldsymbol{\sigma} = \frac{1}{J} \sum_{k} \sum_{\vec{r}_0} S_k \; R_k(\vec{r}_0) \frac{1}{\lambda(\vec{r}_0)} \; f_k^*(\lambda(\vec{r}_0)) \; \boldsymbol{F} \cdot \vec{r}_0 \; \vec{r}_0 \cdot \boldsymbol{F}^c - p\boldsymbol{I} \tag{2}$$

with:

- $\boldsymbol{F}$ the deformation tensor and $J = det(\boldsymbol{F})$, a measure for the volume strain,
- $\vec{r}_0$ a unit vector tangent to the fibre in reference configuration,
- $\lambda(\vec{r}_0)$ the elongation ratio of the fibres oriented in the direction $\vec{r}_0$ in the reference state,
- S_k the volumetric fraction (out of the total volume) of fibres of type k in the unstrained state,
- $R_k(\vec{r}_0)$ the fraction of all fibres of type k oriented in direction $\vec{r}_0$ in the reference state; it should hold that $\sum_{\vec{r}_0} R_k(\vec{r}_0) = 1$,
- $f_k^*(\lambda(\vec{r}_0))$, the load per unit undeformed cross-sectional area, which also accounts for the undulation of the fibres, according to a normal distribution $N(\mu, \sigma)$.)

This model is implemented (Feron, 1993) in the finite element code MARC (MARC, 1994).

4. Identification technique

Hendriks (Hendriks, 1991) presented an identification technique for the characterization of the mechanical behaviour of complex materials. In this method it is assumed that a sufficiently accurate constitutive model is available and its purpose is to estimate the material parameters in these constitutive equations. In contrast with standard methods, the approach no longer demands a homogeneous stress and strain field in some part of the loaded specimen under consideration. Even more, it is preferable to obtain an inhomogeneous strain field (Hendriks, 1991), (Ratingen, 1994).

This numerical-experimental technique is based on the use of three elements (Figure 1):

- measurements of field data on a specimen with arbitrary geometry and boundary conditions,
- finite element modelling of the experiment,
- an iterative scheme to obtain an estimate for the material parameters in the constitutive model employed.

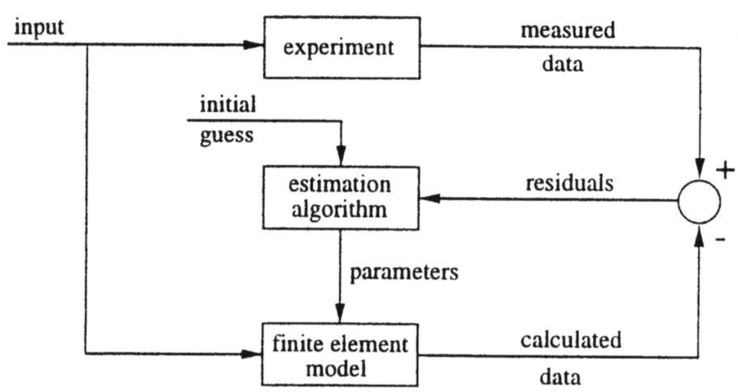

Figure 1: Scheme for the identification technique.

In the following sections the three elements of this numerical-experimental technique are described.

4.1. MEASUREMENT OF FIELD DATA

A common experimental set-up in order to measure (inhomogeneous) strain fields is to place a large number of markers on the surface of the specimen. In the present case a regular ($6 \times 12, 10 \times 22$ mm^2) marker pattern is applied to the surface of the skin using a template and white oil-paint. The positions of the markers are determined with a standard CCD video camera and an image processing and analysis system to obtain the field data. A small surface tensile device to create different states of deformation is used. The pads of this device (see Figure 2) are glued to the skin by means of cyanoacrylate resin, and are moved apart in a number of steps in the experiment (Note that as viscoelastic effects are ignored only quasi-static deformations are considered). See for further details (Voorden, 1996).

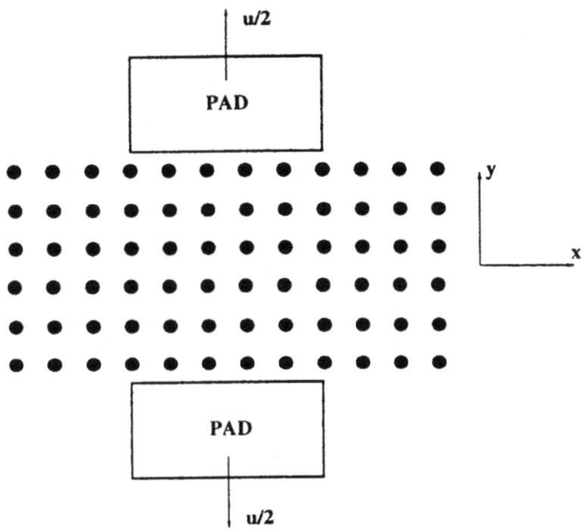

Figure 2: Schematic representation of the camera view: pads and marker pattern.

The advantage of this method is that a large amount of information can be obtained. However, a disadvantage of this method is that it is only possible to measure displacements on the outer surface of the specimen. This means that strain field measurement is only relevant when the surface strain field contains enough information for a characterization of the whole.

4.2. FINITE ELEMENT MODELLING

The next part in the mixed numerical-experimental method is a finite element model of the experiment. For this purpose the finite element code MARC (MARC, 1994) is used. Because of the *in vivo* measurements, the so-called *local approach* is used (Hendriks, 1991), solely using kinematic boundary conditions. Here, only a part of the skin is modelled and a selected set of markers are used to define the edges of the part under consideration (boundary markers). The displacements of these markers are then used as kinematic boundary conditions for the finite element model. As a consequence it may be clear that because forces are not part of the boundary conditions and the measured data, the stiffness parameters cannot be determined absolutely. The finite element calculations can only be carried out for a given set of parameters, thus an initial set of parameters must be available.

In the present case the nodes of the finite element mesh are defined by the measured marker coordinates in the initial state. This is a pragmatic choice, but effective in case of simple geometries. Only the parameters describing the distribution of fibre orientations are estimated. The other parameters are taken from literature.

4.3. PARAMETER IDENTIFICATION

The essential part of this numerical-experimental technique is the iterative procedure that is used to adjust the parameters of the constitutive model employed. The simulation of the experiment results in a set of calculated data which can be compared with a set of data measured in the experiment. The difference (residual) between the two sets contains information about the modelling error, which can be used in order to adjust the parameters. The algorithm employed in this study is essentially different compared to the one presented in the original work (Hendriks, 1991), based on a sequential minimum-variance estimator. The present algorithm is based on a constrained sequential maximum-likelihood approach. The solution procedure and sequential estimator are treated more separately, resulting in a sequential maximum-likelihood estimator, combined with a minimization procedure suited for constrained problems. This minimization procedure takes account for possible equality and inequality constraints as well as possible linear dependencies of the parameters. The algorithm employed will be outlined shortly in the remainder part of this section.

Assume that *a priori* information on the previous (k^{th}) estimate for the parameters, $\underset{\sim}{\hat{x}}_k$, is available and can be modelled by:

$$\underset{\sim}{\hat{x}}_k = \underset{\sim}{x}_t + \underset{\sim}{w}_k, \tag{3}$$

with $\underset{\sim}{x}_t$ a column with the true parameter values and $\underset{\sim}{w}_k$ the estimation error with zero mean and covariance matrix $\underline{P}_k$:

$$E(\underset{\sim}{w}_k) = 0, \quad E(\underset{\sim}{w}_k \underset{\sim}{w}_k^T) = \underline{P}_k, \tag{4}$$

with E the expected value operator. The nonlinear model for the measured or observational data is represented by:

$$\underset{\sim}{y}_{k+1} = \underset{\sim}{h}_{k+1}(\underset{\sim}{x}_t) + \underset{\sim}{v}_{k+1}. \tag{5}$$

The nonlinear function $\underset{\sim}{h}_{k+1}(\underset{\sim}{x}_t)$ describes the dependence of observation $k+1$ on column $\underset{\sim}{x}_t$, if there are no observation errors. Function $\underset{\sim}{h}_{k+1}(\underset{\sim}{x}_t)$

symbolizes the finite element calculation with use of material parameters in column $\underset{\sim}{x}_t$. Column $\underset{\sim}{y}_{k+1}$ contains the observational data from the experiment. Column $\underset{\sim}{v}_{k+1}$ contains observational errors with supposed zero mean and covariance matrix $\underline{R}_{k+1}$. Further it is assumed that $\underset{\sim}{w}_k$ and $\underset{\sim}{v}_{k+1}$ are uncorrelated.

Besides equations (3) and (5), the parameters have to obey a set of equality and inequality constraints,

$$b(\underset{\sim}{x}_t) = 0, \quad c(\underset{\sim}{x}_t) \geq 0. \tag{6}$$

The estimation problem is formulated as an optimization problem and can be written as the following weighted minimization formulation (Starmans, 1994):

$$\min_{\underset{\sim}{\hat{x}}_{k+1}} \{(\underset{\sim}{y}_{k+1} - \underset{\sim}{h}_{k+1}(\underset{\sim}{\hat{x}}_{k+1}))^T \underline{R}_{k+1}^{-1}(\underset{\sim}{y}_{k+1} - \underset{\sim}{h}_{k+1}(\underset{\sim}{\hat{x}}_{k+1}))+$$

$$(\underset{\sim}{\hat{x}}_k - \underset{\sim}{\hat{x}}_{k+1})^T(\underline{P}_k + \underline{Q}_k)^{-1}(\underset{\sim}{\hat{x}}_k - \underset{\sim}{\hat{x}}_{k+1}) \mid b(\underset{\sim}{\hat{x}}_{k+1}) = 0 \wedge c(\underset{\sim}{\hat{x}}_{k+1}) \geq 0\}. \tag{7}$$

It can be derived that matrix $\underline{P}_k$ is updated according to (Starmans, 1994):

$$\underline{P}_{k+1} = \underline{Z}^T[\underline{Z}[\underline{H}_{k+1}^T(\underset{\sim}{\hat{x}}_{k+1})\underline{R}_{k+1}^{-1}\underline{H}_{k+1}(\underset{\sim}{\hat{x}}_{k+1}) + (\underline{P}_k + \underline{Q}_k)^{-1}]\underline{Z}^T]^{-1}\underline{Z}. \tag{8}$$

The matrix $\underline{H}_{k+1}$ expresses the sensitivity of the model output for the parameter variations and is defined as:

$$\underline{H}_{k+1} = \left(\frac{\partial \underset{\sim}{h}_{k+1}(\underset{\sim}{x})}{\partial \underset{\sim}{x}}\right)_{\underset{\sim}{x}=\underset{\sim}{\hat{x}}_k} \tag{9}$$

The matrix $\underline{Z}$ accounts for the active constraints, *i.e.* the constraints which hold as equality. For further details on this it is referred to (Starmans, 1994). Matrix $\underline{Q}_k$ represents an artificial modelling error covariance which is added in order to assure convergence.

The described estimator is implemented in the program PARFIT (Starmans, 1994), employing the NAG library (NAG, 1993). The implemented algorithm is designed to minimize an arbitrary smooth function subject to constraints. Per estimation step, first the minimization according to equation (7) is performed. Secondly the estimate covariance matrix is updated according to equation (8). In this way a sequence of guesses for the solution, *i.e.* estimates, is generated, converging to the solution of the minimization problem.

5. Parameter estimation from in vivo experiment

The marker pattern of 72 markers is applied to the medial site of the forearm of a test person. Two experiments are performed on the same site, but the tension device and the marker pattern are applied in different directions. The first experiment (longitudinal) is used to estimate model parameters. The second experiment (transverse) is used as verification. Strain levels up to approximately 25% where created.

The implementation of Lanir's Skin Model allows the user to define a discrete fibre density function $R_{k,\vec{r}_0}$. In our case the number of fibre directions is eight. From literature it is known (Finlay, 1969) that the collagen fibres are oriented in a three dimensional but primarily planar wavy array. A reasonable assumption therefore is to model the greater part of the collagen fibres in the xy-plane. For numerical reasons it is required to model some stiffness in all directions. That is why the four mean diagonal directions of the unit cube are modelled. It is assumed that a small part of the collagen fibres (20%) is distributed among these four directions. The elastin fibres are assumed to be distributed isotropically in the skin, *i.e.* equally divided among the four diagonal directions. All other parameters in Lanir's Skin Model are assumed to be known, and set to a fixed value.

To initialize the estimation sequence an initial guess for the parameter values to be estimated, $\hat{\underset{\sim}{x}}_0$, and an initial guess for the parameter covariance matrix $\underline{P}_0$ is needed. The column of parameters to be estimated is defined by: $\underset{\sim}{x} = [R_1\ R_2\ R_3\ R_4]^T$. The initial guess is set to: $\hat{\underset{\sim}{x}}_0 = [0.2\ 0.2\ 0.2\ 0.2]^T$. Further, the parameters have to obey one equality constraint: $\sum_i R_i = 0.8$, $i = 1,.,4$ and four additional inequality constraints based on physical grounds: $R_i \geq 0.0$, $i = 1,..,4$. $R1$ to $R4$ are defined as the direction with an angle with the positive x-axis (see Figure 2) of $0, 90, 45, 135°$ respectively.

All parameters are considered initially to be mutually independent, *i.e.* $\underline{P}_0$ is considered to be diagonal. The initial standard deviations of the parameters to be estimated are set to 10^{-1}. Confidence in the model can be expressed by the model covariance matrix $\underline{Q}$. Again, all model covariances between different parameters are assumed to be zero; $\underline{Q}$ is considered to be diagonal. The choice of $\underline{Q}$ can be arbitrary but the diagonal elements are often set to $\underline{P}_{0_{ii}}/100$ (Ratingen, 1994); the model standard deviations are set to 10^{-2}.

The accuracy of the measured positions of the observation markers can be taken into account in the observational covariance matrix $\underline{R}$. All observational covariances between different observational quantities are assumed to be zero; $\underline{R}$ is considered to be diagonal. Initially, the standard deviations of the observations are set to the values according to the experimentally determined standard deviations of the marker positions:

$\sigma_x = 0.004\ mm,\ \sigma_y = 0.003\ mm.$

The various standard deviations have a more or less physical meaning. However, it may also be advantageous to choose their values based on convergence considerations. Especially, when choosing the model standard deviations too small, this may slow down the convergence enormously (Hendriks, 1991), (Starmans, 1994).

Employing different initial guesses for the parameters resulted in approximately the same estimation results. The estimation results in the following set of parameters: $\underset{\sim}{x} = [R_1\ R_2\ R_3\ R_4]^T = [0.000\ 0.054\ 0.526\ 0.220]^T$.

The second experiment is performed in transverse direction, *i.e.* perpendicular to the longitudinal experiment. The transverse experiment is simulated numerically. The finite element mesh and the kinematic boundary conditions are deduced from the experiment carried out. The FE-analysis is performed employing the final estimates of the longitudinal experiment, taking into account the rotation of 90° of the local coordinate system. The predicted marker positions can be compared with the observed marker positions of the experiment. The sample mean and standard deviation of the residuals (measured minus estimated marker displacements) are more or less the same as those in the longitudinal case.

6. Conclusions

In this paper a mechanical characterization of skin was presented, employing a structural skin model, *i.e.* Lanir's Skin Model. An attempt was made to estimate a limited set of parameters, describing the collagen fibre distribution in the xy-plane. Further, a verification experiment was performed in order to examine the results on consistency. Evaluation of the results leads to the following general conclusions:

- The estimation process is successful: the parameters converge to a stable value and the same parameter values are found for different initial guesses.
- Judging the residuals, the results seem to be influenced by modelling errors; the residuals are lager than expected and the residual fields exhibit structure, rather than randomness.

Modelling errors can be attributed to a number of reasons:

- The assumption that the measured surface is perfectly flat; recognition of slight curvature would be more realistic.
- The assumption that the properties within the modelled area are homogeneous.
- Lanir's Skin Model is restricted to the description of the dermis; the influence of the epidermis and hypodermis is neglected.
- Modelling elastic behaviour; skin exhibits viscoelastic behaviour.

References

Brown, I.A.,
Scanning Electron Microscopy of Human Dermal Fibrous Tissue,
Journal of Anatomy Vol. 72, pp. 159-168, 1972

Douven, L.F.A., Fraij, F., Kosters, E.J.A.,
A procedure to determine the material parameters of the human skin in vivo: verification on rubber,
ISBS-10 symposium, Cincinnati, Ohio, 1994.

Feron, R.P.M.J.,
A 3-D Structural Skin Model: Development and Implementation,
Unclassified Nat.Lab.Report 024/93, 1993.

Finlay, B.,
Scanning Electron Microscopy of he Human Dermis under Uniaxial Strain,
Biomedical Engineering Vol. 4, pp. 322-327, 1969.

Hendriks, M.A.N.,
Identification of the Mechanical Behaviour of Solid Materials,
Ph.D. thesis, Eindhoven University of Technology, The Netherlands, 1991.

Lanir, Y.,
A Structural Theory for the Homogeneous Biaxial Stress-Strain Relationships in Flat Collageneous Tissues,
Journal of Biomechanics Vol. 12, pp. 423-436, 1974.

Lanir, Y.,
Constitutive Equations for Fibrous Connective Tissues,
Journal of Biomechanics Vol. 16, pp. 1-12, 1983.

Manschot, J.F.M.,
The Mechanical Properties of Human Skin In-Vivo,
Ph.D. thesis, Catholic University of Nijmegen, 1985.

MARC Analysis Research Corporation,
Volumes A-E of the MARC Reference Library manuals, revision K.5,
Palo Alto, USA, 1994.

NAG, Numerical Algorithms Group Ltd.,
Version Mark 15 of the NAG Fortran Library Manual,
Oxford, United Kingdom, 1993.

Ratingen, M.R., van,
Mechanical Identification of Inhomogeneous Solids,
Ph.D. thesis, Eindhoven University of Technology, The Netherlands, 1994.

Ridge, M.D. and Wright, V.,
The Directional Effect of Skin - a Bio-engineering Study of Skin with Particular Reference to Langer's Lines,
Journal of Investigative Dermatology Vol. 46, pp. 341-346, 1966.

Savenije, E.P.W.,
Structure and Mechanical Properties of Human Hair and Skin,
Nat.Lab. Report No. 5806, 1982.

Starmans, F.J.M.,
Parfit: a Parameter Estimation Tool,
Nat.Lab Report No. 6712, 1994.

Voorden, W.K.L., van der,
Characterization of the in-plane mechanical behaviour of human skin
MSc. thesis, WFW Report No. 96.073, Eindhoven University of Technology, The Netherlands, 1996.

DETERMINATION OF TRABECULAR BONE TISSUE ELASTIC PROPERTIES BY COMPARISON OF EXPERIMENTAL AND FINITE ELEMENT RESULTS

B. VAN RIETBERGEN, J. KABEL*, A. ODGAARD* AND R. HUISKES
Biomechanics Section, Institute of Orthopaedics, University of Nijmegen, P.O. Box 9101, 6500 HB Nijmegen, The Netherlands
**Orthopaedic Research Laboratory, University Hospital Aarhus Building 1A 1st floor, Nørrebrogade 44, 8000 Aarhus C, Denmark*

1. Introduction

Trabecular bone, found near the ends of long bones and inside vertebral bodies, is a spongy type of bone with a very complex internal architecture build of rods and plates. Its main function is to distribute compressive loads in the skeleton to carticular surfaces. Due to this load carrying function, its mechanical properties are of great importance for the integrity of bones. In particular in the elderly, the integrity of trabecular bone can be jeopardized due to loss of bone mass or deterioration of its internal architecture as a result of bone diseases (osteoporosis) or adverse bone remodeling effects (for example: after long bedrest). In particular in those situations, an accurate assessment of mechanical properties of trabecular bone is of great importance to diagnose the bone fracture risk, such that preventive measures can be taken in time when needed.

The mechanical properties of trabecular bone are dependent on two factors. First the mechanical properties of the bone tissue itself, which are correlated with the degree of mineralization. The second is morphology which represents the internal arrangement of the trabecular architecture. Trabecular architecture can be represented by two factors. First the volume fraction, which determines the volume of bone tissue per unit of volume. This quality may also be expressed in the total mass per unit of bone volume, which is called the apparent density, and comprises the mineralization as well. Second the directionality of the trabecular anisotropy. Strong dependencies of bone mechanical properties (stiffness and strength) on its apparent density have been demonstrated in a large number of studies (Carter and Hayes, 1977; Rice *et al.*, 1988; Hodgskinson and Currey, 1992). More recent studies have also included a measure of the directionality of the architecture to further improve the predictive value of these relationships (Turner *et al.*, 1990; Goldstein *et al.*, 1993; Goulet *et al.*, 1994). Much uncertainty exist, however, with respect to the mechanical properties of the bone tissue material.

In earlier studies the tissue elastic properties and strength have been measured using standard engineering test methods such as tensile tests, three- or four-point bending tests and buckling tests. Values found for the tissue modulus range from 0.76 to 10 GPa when using tensile tests, from 3.2 to 5.4 GPa for three- or four-point bending tests and

H. Sol and C.W.J. Oomens (eds.), Material Identification Using Mixed Numerical Experimental Methods, 183-192.

from 8.7 to 14 GPa for buckling tests (for an overview see Rho *et al.*,1992). A major problems when using standard engineering tests methods for the determination of bone tissue properties is the small size of trabeculae (thickness: 100-200 micron, length 1-2 mm), resulting in inaccuracies in the displacements measurements and thus in the calculation of moduli. Another problem is the irregular shape of trabeculae, where standard engineering tests require a standardized specimen geometry. To overcome the latter problem, some studies have used machined specimens (Choi *et al.*, 1990; Choi and Goldstein, 1992), but it is unclear to what extent machining artifacts can affect the stiffness of the specimens. A somewhat less common engineering test method that has been used for the measurement of tissue elastic properties is ultrasound measurement. Values found with this method are generally higher than those obtained from the standard tests: 11 to 15 GPa (Ashman and Rho, 1988; Rho *et al.*, 1992). Recently, nanoindentation test have been developed for the measurement of tissue elastic properties (Rho *et al.*, 1997), providing a tissue modulus at the microlevel ranging from 15-20 GPa.

Based on the large variation in the results obtained from these studies, it has been questioned if *the* tissue properties of trabecular bone can be defined at all. Bone tissue is a structure at all levels of organization (Reilly and Burstein, 1974) and, consequently, different properties will be found if different test methods are performed at different levels. It has been proposed that each test will measure 'a stiffness' for some level of organization rather than an intrinsic property of bone material (Choi *et al.*,1990). The obvious question then is: how to measure tissue properties that are relevant for the mechanical behavior of trabecular bone at the apparent (global) level?

To answer this question, an indirect way for the determination of tissue material properties has been proposed. With this approach, the results of experimental tests for larger bone specimens are compared with those calculated from computer models, based on the finite element method, that represent the internal architecture of the tested specimens. By adjusting the tissue modulus in the finite element model such that experimental and simulation results coincide, a tissue modulus is found that represents the relevant properties *in situ*. Earlier studies (Williams and Lewis, 1982; Gibson, 1985), have used two-dimensional finite element models and finite element models that represent idealized bone architectures, modeled as a repetitive structure build of unit cells. In general, however, trabecular bone architecture can not be represented as such. More recently, three-dimensional reconstruction techniques in combination with large-scale finite element models have been introduced, that allow for finite element modeling of the trabecular architecture of reconstructed specimens in full detail (Fyhrie *et al.*, 1992; Hollister *et al.*, 1993;Van Rietbergen *et al.*, 1995a). In earlier studies we have demonstrated that, using this indirect approach, it is possible to determine an 'effective' isotropic modulus for the tissue material of a reasonable large specimen (1 cm^3) (Van Rietbergen *et al.*, 1995b).

The purpose of this chapter is to demonstrate that the 'effective' isotropic modulus calculated from this combined experimental – large-scale-FE approach indeed is a relevant modulus in the sense that it can be used to predict the apparent properties of trabecular bone specimens and in the sense that it can be considered as an intrinsic material property of trabecular bone tissue *in situ*, that can be determined in an accurate and reproducible way. In the following, the combined experimental – large-scale FE-method for the determination of tissue elastic properties will be described first. An assessment of factors that can affect the results will be given. The effects of boundary artifacts in compression tests is further investigated, since this has been described as one of the most common and critical errors. The reproducibility is then determined by calculating the effective modulus for four trabecular bone specimens that originate from the same bone.

2. The Combined Experimental – Numerical Approach using Large-scale FE-models of Trabecular Bone

With the large-scale FE-techniques, the internal architecture of trabecular bone specimens is represented in detail by a large number of elements. The complex internal geometry of these models is based on high-resolution three-dimensional computer reconstruction techniques. Presently, two methods are available for creating such reconstructions. With the first technique, the specimen is embedded in a plastic raisin and placed in an automated microtome to create sequential slices. A digital camera, mounted on top of the microtome, is used to capture the exposed surface after each slice is taken (Odgaard *et al.*, 1994). By stacking the binarized set of images in a computer, a three-dimensional reconstruction is obtained build of voxels (3-D pixels) that can represent either bone or void. The resolution of this technique is on the order of 1/1000 of the specimen size, i.e. 10 microns for a 1 cm specimen. The second technique is a non destructive one that uses sequential cross-sectional images obtained from CT or MRI scanning (Feldkamp *et al.*, 1989; Rüegsegger *et al.*, 1996; Majumdar *et al.* 1997). The resolution of this technique is dependent on the scanning volume. Typically, laboratory micro-CT and micro-MRI scanners can obtain a resolution of 15 microns for a 15 mm specimen size.

These three-dimensional computer reconstructions can be used for the generation of large-scale finite element models by simply converting each voxel that represents bone tissue to an equally sized brick element in a FE-model (Van Rietbergen *et al.*, 1995). Typically, such FE-models will consist of on the order of 1E6 elements for a 10 mm bone specimen. Problems of this size can not be solved with standard FE-solvers. However, by using the special features of these FE-models (all elements have the same size, geometry and orientation) very fast special purpose FE-solvers can be used that can reduce the cpu-time needed for solving to an hour or less when using supercomputers for the calculations (Van Rietbergen *et al.*, 1996a).

By choosing appropriate boundary conditions, such large-scale FE-models can be used to simulate (multi-axial) compression test experiments on the bone specimen. The element material properties in these models represent the bone tissue material properties, which are not known. However, by assuming linear elastic, homogenous and isotropic tissue material behavior, the results of the compression test simulation are a linear function of the tissue Young's modulus. By assuming a reasonable value for this modulus first, the (multi-axial) apparent elastic properties of the specimen as a whole calculated from the FE-simulations can be scaled later to obtain the best fit between the results of the simulations and those of the experiments (Van Rietbergen *et al.*, 1995b). The best fit can be defined as the tissue modulus scaling factor s that minimizes an error function:

$$Err=\sqrt{\sum_{i=1}^{3}\left(E_i^{\exp}-sE_i^{FE}\right)^2}, \tag{1}$$

with $E_i^{\exp}$ the three orthogonal elastic moduli measured in the experiment, and E_i^{FE} the same moduli calculated from the FE-analyses. With this error function, the best fitting tissue modulus E^T can be calculated from:

$$E^T=s_{\min}E_0^T=\frac{\sum_{i=1}^{3}\left(E_i^{\exp}E_i^{FE}\right)}{\sum_{i=1}^{3}\left(E_i^{FE}\right)^2}E_0^T, \tag{2}$$

with E_0^T the tissue modulus used for the FE-analyses.
An overview of the voxel conversion method is shown in Fig. 1.

3. Evaluation of factors that can affect the accuracy and reproducibility of the combined method

The accuracy of the tissue modulus determined by comparing the apparent elastic properties measured from multi-axial experiments and from large-scale FE-simulation of these experiments can be affected by several factors which can be classified in factors related to the experiment, factors related to the reconstruction, factors related to the FE-calculations and factors related to the protocol used.

Factors that can affect the accuracy of compression tests have been well

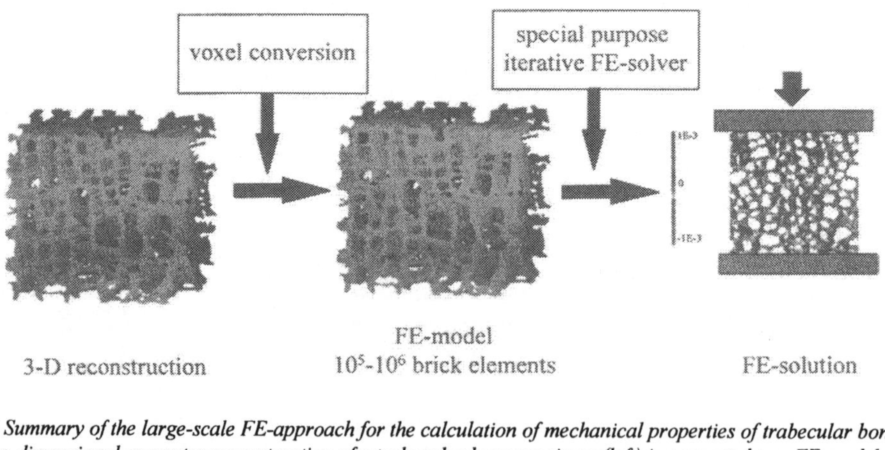

Fig. 1 Summary of the large-scale FE-approach for the calculation of mechanical properties of trabecular bone. A three-dimensional computer reconstruction of a trabecular bone specimen (left) is converted to a FE-model by simply converting all bone voxels in the reconstruction to equally sized brick elements in the FE-model. The resulting FE-model (middle) has exactly the same geometry as the reconstruction it is based on. For a 10 mm specimen, the FE-model consists of on the order of 10^5 to 10^6 elements. To solve these large FE-problems, special-purpose iterative solvers are used. By applying the appropriate boundary conditions, the FE-models can be used to simulate compression test experiments and, after solving the corresponding FE-problem, to calculate the elastic properties of the specimen according to eq. (2).

described (for an overview see Linde, 1993). In these studies, boundary artifacts were identified as one of the most common and critical errors in compression test experiments. Boundary artifacts are created when the bone specimen is excised from its environment. Trabeculae near the sectioned faces of the specimen, which were connected *in situ*, become free-ends, thus forming a region of reduced elastic properties where trabeculae bend and slide. It has been demonstrated that the effect of this artifact can affect the results by as much as 75%, depending on specimen size and geometry and boundary conditions (Keaveny *et al.*, 1993; Odgaard and Linde, 1991). Another factor that can significantly affect the results of compression tests is friction at the bone platen interface. It has been described that, with standard low-friction compression tests, friction can still result in a significant overestimation of the measured Young's modulus (Odgaard and Linde, 1991).

The accuracy of three-dimensional computer reconstructions of trabecular bone is largely determined by the resolution and the slice thickness of the images. In order to image individual trabeculae, the voxel size must be less than the intertrabecular width i.e. on the order of 200 microns. Depending on the purpose of the reconstruction, a (much) smaller voxel can be needed. For the determination of typical morphological parameters (volume fraction, trabecular number, thickness and spacing) it was found that the resolution should be at least 175 microns (Müller *et al.*, 1996). Another parameter that can have a large effect on the resulting reconstruction is the threshold value, used to generate the binary images from the original gray-level images. In particular when using images with a poor resolution, the choice of this parameter can result in a significant thickening or thinning of trabeculae.

Since the geometry of the large-scale FE-models is generated directly from the three-dimensional computer reconstructions, its accuracy depends on the accuracy of the reconstructions as well. The effect of the resolution and consequent element size on the calculation of the apparent elastic properties of a bone specimen was investigated in an earlier study (Van Rietbergen *et al.*, 1995a). In this study it was found that the maximum difference between a model built of 100 micron elements and one build of 20 micron elements was less than 20%, and that this difference was correlated with small differences in volume fraction between the models. The effect of the threshold value on the elastic properties has not been rigorously established but, since the stiffness of the specimen is largely dependent on the volume fraction, it is expected that small changes in the threshold value and, consequently the small changes in the volume fraction of a specimen can have a significant effect on the calculation of elastic properties. Numerical errors in the stiffness calculation can be introduced with the direct conversion of voxels to brick elements, since with this method the surface of trabeculae is not smooth but represented by a 'jagged' composition of brick elements. It has been shown that errors in the stress/strain calculation near the trabecular surfaces can result (Jacobs *et al.*, 1993; Guldberg and Hollister, 1994). However, since these are local artifacts, these errors do not affect the calculation of the stiffness of the specimen as a whole. Other factors that affect the accuracy of the FE-simulation include the fact that the simulation can not capture non-linear phenomena that play a role in the real experiment. For example: the 'toe' that is usually seen in experimental force-displacement curves due to the fact that at low load magnitudes not all trabeculae are load carrying, is not present in the linear FE-models where care is taken that all trabeculae are load carrying. Finally, the modeling of the tissue material properties as linear elastic, homogenous and isotropic can cause differences between experimental and FE-results.

Protocol differences can also be a factor that affects the accuracy of the tissue modulus determined. Differences in storage, and test conditions of the specimen, differences in specimen geometry and size and differences in strain-rate used with testing can result in incomparable results. Consequently, results between experiments can only be compared when using the same protocol.

4. The Effect of Boundary Artifacts on the Determination of the Tissue Modulus

Among all factors that can affect the accuracy of the tissue modulus determination as mentioned in the previous paragraph, those related to the compression test boundary artifacts have been reported as the most common and significant. To address the effect of this artifact for the value of the tissue modulus with the specimen size and test protocol used here, we have developed a special compression test, adapted from Linde and Hvid, (1989), in which the effect of boundary artifacts could be largely excluded (Van Rietbergen *et al.*, 1996b). With this compression test, steel discs are glued to the loaded faces of a cubic specimen of trabecular bone, such that the sliding of trabeculae during testing was restrained. After removing the discs with a blade saw, the test can be repeated with discs glued at the other faces to measure the stiffness in all three orthogonal directions. Since the uniaxial stress condition, required for compression test experiments, is not satisfied in this experiments the moduli measured from this experiment can not be directly compared to those obtained from standard compression tests. However, by simulating the special fixed-platen compression tests in the computer, using a large-scale FE-model that represents the tested specimen and the appropriate fixed-platen boundary conditions, a tissue Young's modulus can be calculated in the same way as described for the standard compression test, which is not affected by boundary artifacts. By comparing the tissue modulus thus obtained to that obtained by comparing experimental and simulation results for a standard compression test for the same specimen, it is possible to quantify the effect of boundary artifacts on the tissue modulus determined.

In an earlier study, a 10 mm specimen taken from the human proximal tibia, was subjected to this special fixed-platen experiment first and a to standard low-friction compression test later (Van Rietbergen *et al.*, 1996b). Due to the fixed boundary conditions, the moduli found in the first experiment (Average= 482 MPa) were higher than those found in the standard test (Average=436 MPa) (Fig. 2). After reconstruction and conversion to a FE-model, both experiments were simulated and the moduli calculated. By fitting the measured and calculated moduli according to eq. (2), a tissue Young's modulus of 14.6 GPa was calculated for the fixed platen experiment and 14.2 GPa for the standard test. With these values, the anisotropic material properties of the specimen as measured in the experiments were well reproduced by the FE-models (Fig. 2), although the fit was somewhat better for the fixed platen experiment (Err= 39.1 MPa) than for the standard test (Err= 74.4 MPa).

Since the difference between the elastic tissue modulus calculated from the special fixed-platen experiment and that calculated from a standard compression test was

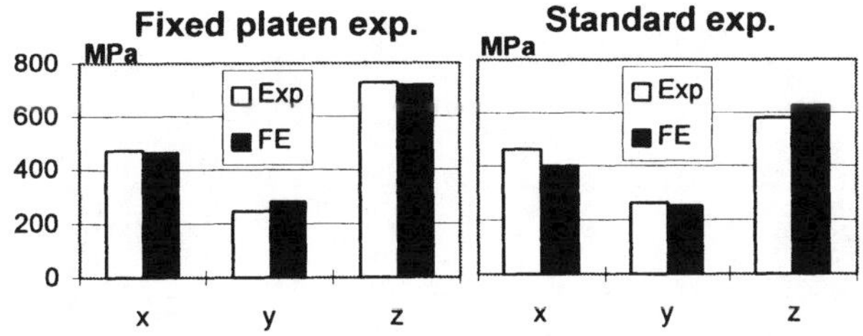

Fig. 2 Apparent Young's moduli measured in the experiments and calculated from the FE-models for the fixed-platen experiment in which the specimen was glued to steel discs (left) and for a standard low-friction compression test (right). The results of the FE-analyses are scaled for the tissue modulus calculated from eq. (2).

less than 3%, it was concluded that, for the specimen size and protocol used here, boundary artifacts would not significantly affect the results of the determined tissue modulus. It is also concluded that the tissue modulus determined for this specimen is relevant in the sense that, in combination with the large-scale FE-models, it can be used to calculate the apparent anisotropic material properties of the bone specimen.

5. Reproducibility: Determination of Tissue Elastic Properties for Four Specimens

To evaluate the reproducibility of the combined experimental–FE approach for the determination of tissue elastic properties, another four specimens were analyzed. All specimens were obtained from one whale vertebral body to ensure that all specimens would have similar tissue properties as much as possible. The specimens were subjected to a standard low-friction compression test while measuring the forces and transversal displacements. From these measurements, three orthogonal moduli and six Poisson's ratios were calculated for each specimen. After testing, three-dimensional computer reconstructions of all specimens were made using the serial sectioning technique (Odgaard *et al.*, 1994). The reconstructions were converted to large-scale FE-models of up to 411,474 elements, and the compression test experiments were simulated with no-friction boundary conditions. Elastic moduli and Poisson's ratio were calculated from the results of the FE-simulations in the same way as done for the real experiments. For each of the four specimens a tissue effective isotropic Young's modulus was calculated that gave the best fit between its three orthogonal moduli determined from the experiment and from the FE-calculations according to eq. (2).

For all specimens the three Young's moduli as measured in the experiments were accurately reproduced by the FE-models. Errors calculated from eq. (1) ranged from 3.2 MPa to 46 MPa for an average modulus of 262 MPa and 307 MPa respectively. Very similar values were found for the tissue modulus of the four specimens, ranging from 5.01 GPa to 5.71 GPa. When plotting the Young's moduli and Poisson's ratios measured in the experiment versus those calculated from the FE-models scaled for the best fitting tissue modulus, it can be seen that there is an excellent correlation between the moduli, and a

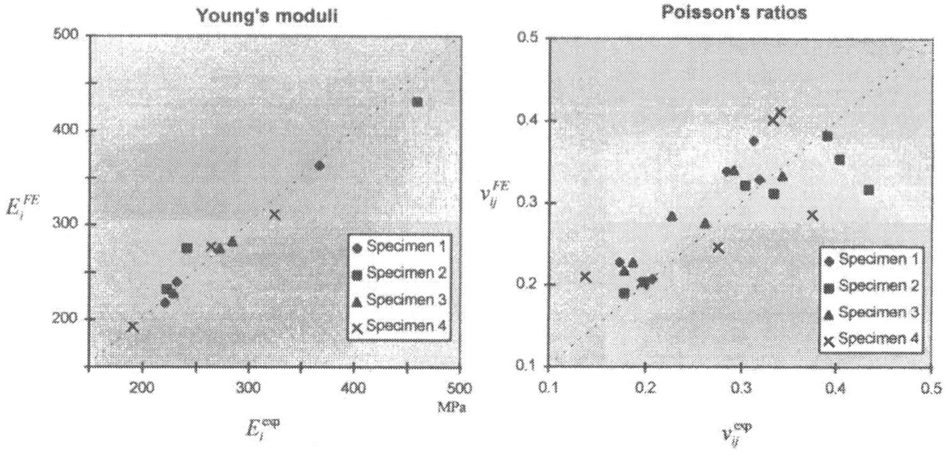

Fig. 3 Young's moduli (left) and Poisson's ratios (right) of the four specimens measured in the experiment versus those calculated from the FE-analyses

good correlation between Poisson's ratios (Fig. 3). The somewhat lesser correlation for the Poisson's ratios is probably due to the larger errors in the measurements of transversal displacement compared to those of the longitudinal displacements in the experiments and due to small differences between these measurements in the experiment and in the FE-model.

Since the variation in the tissue modulus determined for four similar specimen was only 13%, it was concluded that the calculated tissue modulus represents an inherent material property of the trabecular tissue that can be determined in an accurate and reproducible way. Part of the remaining variation in the tissue modulus can be due to variation in the tissue material properties of the specimens. It is possible that the variation is further reduced if more information about the tissue properties is available (in particular the tissue mineralization).

6. Discussion

For all specimens investigated, the anisotropic material properties are well represented by the FE-model with an isotropic 'effective' tissue modulus. It was thus concluded that the 'effective' tissue modulus is a relevant parameter in the sense that, in combination with these large-scale FE-models, it can be used to calculate the anisotropic material properties of trabecular bone. It was also found that the effective tissue moduli determined for specimens that originate from the same bone vary only within a narrow range, and thus, that the tissue modulus is relevant in the sense that it is accurate and reproducible.

The tissue modulus found for the human specimen (14.6 GPa) is much larger than the average value found for the whale bone specimens (5.6 GPa). These differences can not be explained by differences in the level of testing, since the same test method was used in both cases and since both types of bone have similar morphological properties.

Consequently, these differences must be due to differences in the mineral content and microstructure of the bone tissue. As indicated for the whale bone specimens, differences in tissue mineralization can explain a part of the variation in the tissue modulus. It has been proposed, however, that tissue density alone might not be a sufficient estimator for modulus when comparing structural different bone, but that microstructural variations should be accounted for (Choi and Goldstein, 1992). To determine a realistic range for the variation of tissue moduli between species, individuals and locations, a much larger and more diverse set of specimens will be needed.

When considering the microarchitecture of trabecular tissue material, the question may arise how it is possible that such good results are obtained when using models with isotropic tissue properties. The tissue material consists of a plywood-like lamellar organization of a collagen-crystal composite, with lamellae aligned with the trabeculae. Consequently, it is unlikely that the tissue material properties are isotropic. However, since most trabeculae are loaded in bending or compression, the longitudinal modulus of the trabeculae largely determines the mechanical behavior of the tissue *in situ*. It is thus likely that the 'effective' isotropic tissue modulus found with the methods described in this chapter in fact represents the longitudinal stiffness of trabeculae, whereas their stiffness in transversal directions can be less.

An important consequence of the finding that the anisotropic material properties are well predicted from FE-models with isotropic tissue properties is that the anisotropy of trabecular bone must be purely due to its architecture. This implies that relationships to predict the anisotropic material properties of trabecular bone from measurements of its density and its morphology only can exist.

Acknowledgments This project was sponsored by The Netherlands Foundation for Research (NWO/Medical Sciences), by the National Computing Facilities Foundation (NCF), by the Danish Rheumatism Association (233-588) and by the Danish Research Councils (5210563-1 and 9313411).

References

Ashman R.B. and Rho, J.Y. (1988) Elastic modulus of trabecular bone material, *J. Biomech.*, **21**, 177-181

Carter D. R. and Hayes, W.C. (1977) The behavior of bone as a two-phase porous structure, *J. Bone Jt. Surg.* **59-A**, 954-962

Choi, K., Kuhn, J.L., Ciarelli, M.J. and Goldstein, S.A. (1990) The elastic moduli of human subchondral, trabecular and cortical bone tissue and the size-dependency of cortical bone modulus, *J. Biomech.*, **23**, 1103-1113

Choi, K. and Goldstein, S.A. (1992) A comparison of the fatigue behavior of human trabecular and cortical bone tissue, *J. Biomech.*, **25**, 1371-1381

Feldkamp, L.A., Goldstein, S.A., Parfitt, A.M., Jesion, G. and Kleerekoper, M. (1989) The direct examination of three dimensional bone architecture in vitro by computed tomography, *J. Bone Min. Res.*, **4**, 3-11

Fyhrie, D.P., Hamid, M.S., Kuo, R.F. and Lang, S.M. (1992) Direct three-dimensional finite element analysis of human vertebral cancellous bone, *Proc. 38th Meeting ORS*, p. 551

Gibson L.J. (1985) The mechanical behavior of cancellous bone, *J. Biomech.*, **18**, 317-328

Goldstein S.A., Goulet, R. and McCubbrey, D. (1993) Measurement and significance of three dimensional architecture to the mechanical integrity of trabecular bone, *Calcif. Tiss. Int.* **53**, S127-133

Goulet, R.W., Goldstein, S.A., Ciarelli, M.J., Kuhn, J.L., Brown, M.B. and Feldkamp, L.A. (1994) The relationship between the structural and orthogonal compressive properties of trabecular bone, *J. Biomech.*, **27**, 375-389

Guldberg, R.E. and Hollister, S.J. (1994) Finite Element solution errors associated with digital image-based mesh generation, *ASME-BED*, **28**, 147-148

Hodgskinson, R. and Currey, J.D. (1992) Young's modulus, density and material properties in cancellous bone over a large density range, *J. of Mat. Science: Materials in Medicine*, **3**, p. 377-381

Hollister, S.J., Brennan, J.M. and Kikuchi, N. (1994) A homogenization sampling procedure for calculating trabecular bone effective stiffness and tissue level stress, *J. Biomech.*, **27** 433-444

Jacobs, C.R., Mandell, J.A. and Beaupré, G.S. (1993) A comparative study of automatic Finite Element mesh generation techniques in orthopaedic biomechanics, *ASME-BED*, **24**, 512-514

Keaveny, T.M., Borchers, R.E., Gibson, L.J. and Hayes, W.C. (1993) Theoretical analysis of the experimental artifact in trabecular bone compressive modulus, *J. Biomech.*, **26**, 599-607

Linde, F. (1993) Elastic and viscoelastic properties of trabecular bone by a compression testing approach. Laegeforeningens Forlag, Kobenhavn

Linde, F. and Hvid, I. (1989) The effect of constraint on the mechanical behaviour of trabecular bone specimens, *J. Biomech.*, **22**, 485-490

Majumdar , S., Newitt, D.C., Link, T.M., Augat, P., Lin, J.C., Ouyang, X. Lang, T. Lotz, J. and Genant, H.K. (1997) Three dimensional assessment of trabecular architecture using magnetic resonance imaging, *Proc. 43rd Meeting ORS*, p. 209

Müller, R., Koller, B., Hildebrand, T. Laib, A., Gianolini, S. and Rüegsegger, P. (1996) Resolution dependency of microstructural properties of cancellous bone based on three-dimensional μ-tomography, *Technol. Health-Care*, **4**, 113-119

Odgaard, A., Andersen, K., Ullerup, R., Frich, L.H. and Melsen, F. (1994) Three-dimensional reconstruction of entire vertebral bodies, *Bone*, **15**, 335-342

Reilly D.T. and Burstein, A.H. (1974) The mechanical properties of cortical bone, *J. Bone Jt. Surg.*, **56A**, 1001-1022.

Rho, J.Y., Ashman, R.B. and Turner, C.H. (1993) Young's modulus of trabecular and cortical bone material: ultrasound and microtensile measurements, *J. Biomech.*, **26**, 111-119

Rho, J.Y., Roy, M.E., Tsui, T.Y. and Pharr, G.M. (1997) Young's modulus and hardness of trabecular and cortical bone in the various directions determined by nanoindentation, *Proc. 43rd Ann. Meeting ORS*, p. 811

Rice, J.C., Cowin, S.C. and Bowman, J.A. (1988) On the dependence of the elasticity and strength of cancellous bone on apparent density, *J. Biomech.*, **21**, p. 155-168

Rüegsegger, P., Koller, B. and Müller, R. (1996) A microtomographic system for the nondestructive evaluation of bone architecture, *Calc. Tiss. Int.* **58**, 24-29

Turner C.H., Cowin, S.C., Rho, J.Y., Ashman, R.B. and Rice, J.C. (1990) The fabric dependence of the orthotropic elastic constants of cancellous bone. *J. Biomech.* **23**, 549-561

Van Rietbergen, B., Weinans, H., Huiskes, R. and Odgaard, A. (1995a) A new method to determine trabecular bone elastic properties and loading using micromechanical Finite-Element models, *J.Biomech.*, **28**, 69-81

Van Rietbergen, B., Huiskes, R., Weinans, H., Odgaard, A. and Kabel, J. (1995b) The role of trabecular architecture in the anisotropic mechanical properties of bone, in: *Bone structure and remodeling* (Ed. by Odgaard, A. and Weinans, H.), World Scientific, Singapore

Van Rietbergen, B. Weinans, H., Polman, B.J.W., and Huiskes, R. (1996a) Computational strategies for iterative solutions of large FEM applications employing voxel data, *Int. J. Num. Meth. in Eng.*, **39**, 2743-2767

Van Rietbergen, B., Odgaard, A. Kabel, J. and Huiskes, R. (1996b) Determination of tissue elastic properties by comparison of experimental and Finite Element results, *Proc. 10th Meeting ESB*, p. 327

Williams J.L. and Lewis J.L. (1982) Properties and an anisotropic model of cancellous bone from the proximal tibia epiphysis, *J. Biomech. Eng.*, **104**, 50-56

MIXED NUMERICAL/EXPERIMENTAL METHOD FOR THE IDENTIFICATION OF DAMAGE IN COMPOSITE MATERIALS

P.J-P. BOUQUET
Vrije Universiteit Brussel, Dienst Analyse van Strukturen
Pleinlaan 2, 1050 Brussel (Belgium)

Abstract. This paper presents a mixed numerical/experimental method for the identification of micro-mechanical damage in composite materials. The stiffness reduction due to matrix cracks is evaluated based on the theory of damage continuum mechanics. The model should be capable to describe the anisotropic post-damage stiffness behaviour and the stress dependant damage evolution in composite materials. The constitutive behaviour of the damaged composite material can be implemented in finite element models. The parameters in the damage model are identified by comparing the results of computations with finite element models of teststructures with experimental results. The experiments include static and dynamic tests on beams and plates.
Different estimators are considered: Bayesian, maximum likelihood and the Kalman filter. A short discussion of the different estimators will be presented.

1. Introduction

A brief summary will be given of a research strategy, based on a mixed numerical/experimental approach that will lead to the identification of damaged composite materials. The aim is to develop a non-destructive technique which allows to identify state variables by means of measurable dynamical parameters. It was H. Sol [1] who first applied such a MNET by developing a numerical model for an anisotropic thin plate, which was implemented in an iterative parameter estimation algorithm. This achievement led to an easy-to-use, non-destructive, stiffness identification method for orthotropic materials, known as the Resonalyser. His work was further elaborated by Hua H. [2] for thick plates, sandwich panels and circular orthotropic disks. J. De Visscher [3] focused her research on the identification of damping properties of thin orthotropic plates which was incorporated in the Resonalyser.
In a MNET two fundamental models are involved: an experimental model and a numerical model. An appropriate estimation algorithm, which is in general iterative, is used to update

H. Sol and C.W.J. Oomens (eds.), Material Identification Using Mixed Numerical Experimental Methods, 193-202.

a set parameters of the numerical model by correlating the static and/or dynamic characteristics of the numerical model with the experimental results. Identification of material properties using vibrational data is a special application of system identification: the configuration of the system and the boundary conditions are well-known, the geometry can be precisely measured and the material properties are the parameters to be updated. In this research damage will be introduced in composite beams and plates by means of periodic vibrational excitation with a shaker. The amplitude will be chosen so that only the brittle matrix of the thin composite plate is damaged and the glassfibre reinforcement remains in a virgin state. Not only the rate of damage will be investigated but also the rate of damage growth during these cyclic damaging sequences.

2. Cementitious fibre reinforced composites

The material which will be considered in this research consists of cementitious, glassfibre reinforced thin plates. The cementitious matrix material, known as IPC, and compatible with glassfibre reinforcements, was recently developed at the laboratory of civil construction at the VUB. Like plain concrete, IPC, is a brittle material with low tensile strength and strain capacities [4]. The main role of the reinforcement is to control the cracking and to alter the behaviour of the material once the matrix has cracked. Thus the fibres improve the ductility of the material and consequently, its energy absorption capacity. Cementitious composites are characterized by a transition zone in the vicinity of the reinforcing inclusion, in which the microstructure of the matrix is considerably different from that of the bulk matrix. It can be shown that the matrix in the vicinity of the fibre is much more porous than the bulk matrix. A distinction should be made between monofilament fibres and bundled filaments (e.g. glassfibre), with the latter only the external filaments tend to have direct access to the matrix. Long glassfibre reinforcements will be used for the fabrication of orthotropic test specimens.

3. Damage and damage growth model

The theory of damage describes the evolution of the phenomena between the virgin state and macroscopic crack initiation. The damage process is not always easily distinguishable from the deformation phenomena which usually acompany it. The one we will focus on is fatigue damage, caused by stress repetitions and identified as a function of the number of cycles. Distinction should be made between the linear and nonlinear accumulation of damage. The Palmgreen-Miner rule is based on the assumption that damage is accumulated additively when it is defined by the associated life ratio $\frac{N_i}{N_{Fi}}$, N_i being the number of cycles applied under a given load for which the the number of cycles to fracture (under periodic conditions) would be N_{fi}. The fracture criterion is:

$$\sum_i \left(\frac{N_i}{N_{Fi}} \right) = 1 \tag{3.1}$$

The definition of a mechanical damage variable itself presents a difficult problem. Lemaitre and Chaboche [5] therefore propose to imagine internal damage variables which are representative of the deteriorated state of the matter. A representative element of finite volume out of a damaged volume is considered. Every area of a section of the volume element, which is identified by its normal, contains cracks and cavities. From this fact we introduce the effective area of resistance $\tilde{S}$, taking account of the area of these traces, stress concentrations in the neighbourhood of geometric discontinuities, and the interactions between the neighbouring defects. Let S_D be the difference between the overall area S and the effective area, the total area of defect traces corrected for stress concentrations and interactions. Because of the lack on information about the value of S_D we will have to introduce the concept of effective stress associated to the hypothesis of strain equivalence. Let us define the mechanical measure of local damage relative to the direction of $\vec{n}$.

$$D_n = \frac{S_D}{S} \tag{3.2}$$

From a physical point of view, the damage variable is therefore the relative area of cracks and cavities cut by the plane normal to the direction $\vec{n}$. Mathematically D_n is the corrected surface density of the discontinuities of the matter in the plane normal to $\vec{n}$. $D_n = 0$ represents the initial non damaged or virgin state, $D_n = 1$ corresponds to the breaking of the volume element into two parts along a plane normal to $\vec{n}$. Any value of D_n in between 0 and 1 characterizes the damaged state. In the general case of anisotropic damage consisting of cracks and cavities with preferred orientations, the value of D_n will depend on the orientation of the normal. Damage is not directly accessible to measurement, its quantitative evaluation is linked to the definition of the variable choosen to represent the phenomenon. The modification of mechanical properties is therefore considered to be caused by damage. The effective stress is defined as the stress calculated over the section which effectively resists the forces. The effective stress, by definition is equal to:

$$\tilde{\sigma} = \frac{\sigma S}{\tilde{S}} = \frac{\sigma}{1 - D} \tag{3.3}$$

in which $\tilde{S}$ the effective area of resistance is defined as $\tilde{S} = S - S_D = S(1 - D)$

We will furthermore assume that the deformation behaviour of the material is only affected by damage in the form of effective stress. The constitutive laws for the virgin material will remain except that the usual stress will be substituted by the effective stress. Once introduced this parameter we will try to determine the damage when measuring the variation of the modulus of elasticity. From

$$\tilde{\sigma} = \frac{\sigma}{1 - D} = E\varepsilon_e \text{ and } \sigma = E(1 - D)\varepsilon_e = \tilde{E}\varepsilon_e \tag{3.4}$$

we find that:

$$D = 1 - \frac{\tilde{E}}{E} \tag{3.5}$$

The physical significance of the effective stress concept is that the stress transformation tensor is a mapping between the mechanical problem in the real physical space (damaged) which can be described in the effective space (virgin state) with the equivalent strain. Thus the problem of continuum damage mechanics can be solved by any technique of nonlinear continuum mechanics in a fictitious space. We will assume a cyclic fatigue damage model to describe the change of material properties of the specimen. Because of relation (3.1), in periodic tests, damage evolution is considered to be linear in that $D = \frac{N}{N_F}$. There are two ways of defining a damage incremental law incorporating the linear accumulation rule. First, following well established rules this increment is introduced per cycle by using simply an equation of the form:

$$\delta D = \frac{\delta N}{N_F(\ldots)} \tag{3.6}$$

Employing a nonlinear evolution of damage we become the second formulation for example

$$\delta D = \frac{(1 - D)^{-k}}{k + 1} \frac{\delta N}{N_F(\ldots)} \tag{3.7}$$

The Palmgreen-Miner linear accumulation law gives good results only for loads for which there is little variation in the amplitude and mean value of the stress.

4. Numerical model

A numerical model for thin plates will be considered, based on the assumptions of Love-Kirchhoff and namely that straight sections normal to the middle plane remain normal and straight during deformation. The normal stress σ_z is small and its direct strain ε_z can be neglected. These postulations allow us to describe the state of deformation by one parameter w(x,y). Assume the dynamic equilibrium equations of a free vibrating structure, neglecting the damping, and considering steady state vibrations $\{z(t)\} = \{\Phi\}e^{i\omega t}$. This yields to an eigenvalue equation to be solved, with eigenvalues ω_i^2 and associated eigenmodes $\{\Phi_i\}$:

$$\omega_i^2 = \frac{\{\Phi_i\}^T[K]\{\Phi_i\}}{\{\Phi_i\}^T[M]\{\Phi_i\}} = \frac{K_i}{M_i} \tag{4.1}$$

[M] representing the mass matrix, [K] the stiffness matrix and K_i, th i-th modal stiffness and M_i the i-th modal mass. A first attempt to localize damage was made by Stubbs [6],[7],[8]. When damage is introduced, consider the mass of the structure unaffected by the damage, only a finite number of local stifness parameters k_j is assumed to be modified by Δk_j due to the amount of damage. The variation of ω_i^2 can be expressed in terms of the variation of global stiffness matrix and

$$\frac{\Delta\omega_i^2}{\omega_i^2} = \frac{\Delta K_i}{K_i} = \frac{\{\Phi_i\}^T[\Delta K]\{\Phi_i\} + 2\{\Phi_i\}^T[K]\{\Delta\Phi_i\}}{\{\Phi_i\}^T[K]\{\Phi_i\}} \tag{4.2}$$

As $\Delta\omega_i^2$ and $\{\Delta\Phi_i\}$ can be determined experimentally but $[\Delta K]$ is unknown. $[\Delta K]$ can be expressed in terms of the variations of the element stiffness matrices $[\Delta K^r]$ per element and approximated by first order Taylor series expansion. Substitution in (4.2) yields to the damage detection equations (cfr.[7],[8],[9]),

$$\frac{\Delta\omega_i^2}{\omega_i^2} - \frac{2\{\Phi_i\}^T[K]\{\Delta\Phi_i\}}{K_i} = \frac{1}{K_i}\sum_{j=1}^{m}\left(\sum_{r=1}^{ne}\{\Phi_i\}^T[A^r]^T\frac{\partial[K^r]}{\partial k_j}[A^r]\{\Phi_i\}\right)\Delta k_j \tag{4.3}$$

In the case of uniform damage, $\{\Delta\Phi_i\}$ is zero, the second term on the lefthandside of (4.3) disappears. The localization of the damage zones however is a more complicated problem because modal parameters are global and thus structure dependant. The number of equations and the number of local stiffness parameters being related to each other, the amount of information needed for localisation of the damage will quickly lead to intensive computations. A different experimental approach is explained in paragraph 5. For every variation Δk_j , the eigenvalues are recalculated and the righthandside of (4.3) can be determined.

5. Experimental method

The identification of the initial stiffness properties of the beams and plate will be executed with the Resonalyser and followed by a VPI-Sensor scan. The Vibrational Pattern Imager, shortly VPI sensor, is a non-contacting velocity transducer capable of remote measurements of the velocity of a solid surface. The main application domain for the the instrument is vibration measurement on components and structures. Frequencies and vibration patterns of structures are measured and visualised. A scan of the first mode of the vibrating beams will be made and the plates will be scanned at their torsion, breath and saddle mode vibration. The vibrational excitement for the scan of the beams and plates will be executed at small amplitudes, to avoid the introduction of matrix crack damage. For simple structures the righthandside of (4.3) can be calculated analytically. For example, the i-th modal stiffness of a beam can be expressed as:

$$K_i = \int_L EI\left(\frac{\partial^2 y_i}{\partial x^2}\right)^2 dx \tag{5.1}$$

$y_i(x)$ being the i-th modal deformation, E the elasticity modulus and I the moment of inertia of a beam, with length L. The initial stiffness properties identified a mechanical shaker will be used to put a high amplitude frequency excitation on respectively the beam or plate. Matrix cracks will be introduced without damaging the fibres. At regular cycle intervals the sequence of identifying the stiffness properties of respectively the beams or the plate will be performed and a frequency shift will be measured with the Resonalyser. This change will be regarded as the mean overall damage parameter D (cfr. paragraph 3) of a uniformly and homogeneous, damaged testspecimen. To obtain a qualitative image of the damage distribution, a VPI-Sensor scan will be executed at the actual resonance frequencies. The change of local flexion due to the nonhomogeneous damage will be obtained by comparison of the scan of the initial state and the damaged state. Consider a beam (eq.(5.1)), when damage induces a variation ΔE in the j-th element (cfr. paragraph 4) the variation of K_i is

$$\Delta K_i = \int_{x_j}^{x_j+\Delta x_j} \Delta EI\left(\frac{\partial^2 y_i}{\partial x^2}\right)^2 dx = \frac{\Delta E}{E}\int_{x_j}^{x_j+\Delta x_j} EI\left(\frac{\partial^2 y_i}{\partial x^2}\right)^2 dx \tag{5.2}$$

(5.1), (5.2) substituted in (4.2) yields to the righthandside of (4.3) and can be compared to the numerical solution.

6. Statistical parameter estimation techniques

Assume the general problem in which one tries to estimate an unknown parameter or state variable of a system based on noise corrupted output signals or observations. The value assigned by the estimator to the state variable is called the estimate. An optimal estimation is an assignment of the estimate in accordance with the minimisation of some estimation criterion. The solution of the numerical model is obtained from an eigenvalue equation (4.1). By truncation of the Taylor series expansion in the vicinity of the initial parameter estimate $\{p_0\}$, the relation between the eigenvalues $\{\lambda\}$ and the parameters, can be expressed as a state equation

$$\{X\} = [I]\{X\} \tag{6.1}$$

$$\{Y\} = [T]\{X\} + \{\varepsilon\} \tag{6.2}$$

With the state and observation parameters

$$\{X\} = (\{p\} - \{p_0\}) \text{ and } \{Y\} = (\{\lambda\} - \{\lambda_0\}) \tag{6.3}$$

and [T], the sensitivity matrix defined as $\left[\dfrac{\partial \lambda_m}{\partial p_n}\right]_{p=p_0}$ (6.4)

$\{\varepsilon\}$ the measurement noise, assumed to be white Gaussian, normal distributed with zero mean, covariance matrix $[S_{\varepsilon\varepsilon}]$ and uncorrelated with $\{X\}$.

6.1 UNCONDITIONAL MAXIMUM LIKELIHOOD

Starting from previously mentioned relations and

$$\{\lambda_e\} = \{\lambda\} + \{\varepsilon\} \tag{6.5}$$

$$\{Y\} = (\{\lambda_e\} - \{\lambda_0\}) \tag{6.6}$$

$\{\lambda\}$ is the vector of eigenvalues, $\{\lambda_e\}$ is the observation vector, $\{\lambda_0\}$ is the vector of eigenvalues given the initial parameter $\{p_0\}$ and $\{p\}$ is the parameter vector to be estimated.

Given {Y} and {X} both normally distributed and zero mean, the joint probability density function of the observation vector {Y} will be $p(X,Y) = p(Y|X)p(X)$. From the definition of normal distributions the unconditional maximum likelyhood function can be found to be

$$l(p) = C\exp -\frac{1}{2}\left(\{Y\}^T[S_{\varepsilon\varepsilon}]^{-1}\{Y\} + \{X\}^T[S_{XX}]^{-1}\{X\}\right) \qquad (6.7)$$

C depending on $[S_{XX}]$ and $[S_{\varepsilon\varepsilon}]$ but independent of {p}. The maximisation can be achieved by differentation of the log l(p) (6.7) or the minimisation of an error functional err(p).

$$\min_p \mathrm{err}(p) = \{Y\}^T[S_{\varepsilon\varepsilon}]^{-1}\{Y\} + \{X\}^T[S_{XX}]^{-1}\{X\} \qquad (6.8)$$

We become, taking into account that $[S_{XX}]$ and $[S_{\varepsilon\varepsilon}]$ are symmetric, after substitution of (6.2), (6.3), and (6.6) into (6.8) and the expected value for {p} noted as $\{\hat{p}\}$

$$\{\hat{p}\} = \{p_0\} + [S_{XX}][T]^T\left([T][S_{XX}][T]^T + [S_{\varepsilon\varepsilon}]\right)^{-1}\left(\{\lambda_c\} - \{\lambda_0\}\right) \qquad (6.9)$$

The error covariance matrix has to be derived from:

$$[\hat{S}_{XX}] = E\left[\{\hat{X} - X\}\{\hat{X} - X\}^T\right] \qquad (6.10)$$

and after some cumbersome calculation we obtain

$$[\hat{S}_{XX}] = \left([S_{XX}]^{-1} + [T]^T[S_{\varepsilon\varepsilon}]^{-1}[T]\right)^{-1} \qquad (6.11)$$

The iteration is initialised given $\{p\}=\{p_0\}$ and $[S_{xx}]=[S_{xx}(o)]$

6.2 DISCRETE KALMAN FILTERING

Consider a discretized linear problem, described in a state equation form, cfr.(6.1)-(6.6) then

$$\{\lambda(k)\} - \{\hat{\lambda}(k|k-1)\} = [T(k)]\left(\{p(k)\} - \{\hat{p}(k|k-1)\}\right) + \{\varepsilon_\lambda(k)\} \qquad (6.12)$$

X(k), Y(k), ε(k) and $\varepsilon_\lambda(k)$ are vectors, representing respectively the state parameter, the measured parameter and the measurement noise at k times the sampling period. Where $\{\lambda(k)\}$ is the observation vector and $\hat{\lambda}(k|k-1)$ is the last estimate. {p(k)} is a vector composed of states or parameters to be estimated and discrete in time. This parameter vector is generally subjected to perturbations, model noise. The measurement noise is, identical to paragraph 6.1, white noise with zero mean and a prior known covariance:

$$E\left[\{\varepsilon_\lambda(k_1)\}\{\varepsilon_\lambda(k_2)\}^T\right] = \left[S^\lambda_{\varepsilon\varepsilon}\right]\delta_{k_1 k_2} \tag{6.13}$$

$$E\left[\{\varepsilon_\lambda(k)\}\right] = 0 \tag{6.14}$$

The model noise is introduced as p(k) and has similar statistical properties as the measurement noise, the expectation E[] gives the covariances and the mean values and δ being the Kronecker symbol.

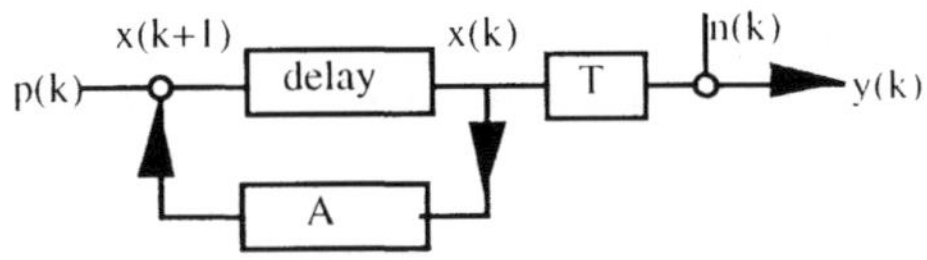

Figure 6.2.1. Iterative procedure of the Kalman filter

If random variables {X} and {Y} are jointly Gaussian, it can be shown from the probability theory that $\left[X^T Y^T\right]^T$ has mean and covariance

$$\begin{Bmatrix}\bar{x}\\ \bar{y}\end{Bmatrix} \text{ and } \begin{bmatrix} S_{XX} & S_{XY}\\ S_{YX} & S_{YY}\end{bmatrix} \tag{6.15}$$

and when the random variable {X} conditional on {Y}=y is Gaussian, then the mean and covariance are respectively

$$\{\bar{x}\} + \left[S_{YX}\right]^T\left[S_{YY}\right]^{-1}\{y - \bar{y}\} \text{ and } \left[S_{XX}\right] - \left[S_{XY}\right]\left[S_{YY}\right]^{-1}\left[S_{YX}\right],\ \left[S_{YY}\right] \tag{6.16}$$

The linear, discrete Kalman Filter for the problem of constant parameter can be derived, with [K] the Kalman gain matrix and I the identity matrix:

$$\left\{\hat{p}(k+1|k)\right\} = \left\{\hat{p}(k|k-1)\right\} + [K]\left[\lambda(k) - \hat{\lambda}(k|k-1)\right] \tag{6.17}$$

$$K = \left[S_{pp}(k|k-1)\right]\left[T(k)\right]^T\left(\left[T(k)\right]\left[S_{pp}(k|k-1)\right]\left[T(k)\right]^T + \left[S_{\varepsilon\varepsilon}(k)\right]\right)^{-1} \tag{6.18}$$

$$\left[S_{pp}(k+1|k)\right] = \left[I - K\left[T(k)\right]\right]\left[S_{pp}(k|k-1)\right] \tag{6.19}$$

Conclusion

As was mentioned in the introduction, this paper proposes a MNET for the identification of micro-mechanical damage in composite material. An experimental measuring technique is explained which allows the measurement parameters to be compared with the results obtained from a numerical damage model. After comparison of the experimental and numerical state parameters, two Bayesian estimators were proposed for the estimation of the state variables. This research proposal should be regarded as the continuation of the expertise obtained at the department of Structural Analysis in the field of MNET and modal analysis.

References

[1] Sol H., *Identification of anisotropic plate rigidities using free vibration data*, Ph. D. thesis, Vrije Universiteit Brussel, 1986.

[2] Hua H., *Identification of plate rigidities of anisotropic rectangular plates, sandwich panels and circular orthotropic disks using vibration data*, Ph.D. thesis, Vrije Universiteit Brussel, 1993.

[3] De Visscher J., *Identification of the complex stiffness matrix of orthotropic materials by a mixed numerical experimental method* , Ph.D. thesis ,Vrije Universiteit Brussel, 1995

[4] Bentur A. and Mindess S., *Fibre reinforced cementitious composites*, Elsevier Applied Science London and New York, ISBN 1-85166-393-2.

[5] Lemaitre J. and Chaboche J-L., *Mechanics of solid materials*, Cambridge University Press, 1994, ISBN 0-521-32853-5

[6] Stubbs N., Osegueda R., *Global non destructive damage evaluation in solids*, The Int.J. of Analytical and Experimental Modal Analysis, Vol.5, n°2, pp. 67-80, 1990.

[7] Stubbs N., Osegueda R., *Global damage detection in solids_Experimental verification*, The Int.J. of Analytical and Experimental Modal Analysis, Vol.5, n°2, pp. 81-98, 1990.

[8] Sanders D.R., Kim Y. I., Stubbs N., *Nondestructive evaluation of damage in composite structures using modal parameters*, Cambridge University Press, 1994, ISBN 0-521-32853-5

INVERSE IDENTIFICATION OF THE INTERNAL LENGTH SCALE OF QUASI-BRITTLE MATERIALS USING NONLOCAL DAMAGE MODELS

J. CARMELIET
Catholic University of Leuven, Department of Civil Engineering, Celestijnenlaan 131, 3001 Heverelee, Belgium

1. Introduction

Failure in quasi-brittle materials involves a progressive damage process, which is initially driven by the growth and coalescence of micro-cracks and finally leads to the localization of deformation in narrow damage zones, such as cracks in concrete and shear bands in soils. Due to this damage process, the load carrying capacity of the material decreases at increasing deformation, which may result in a so-called strain-softening behaviour. When a constitutive law with a softening branch is applied in a standard continuum damage model, the strain softening phenomenon leads at the onset of localization to an ill-posed boundary value problem, since in quasi-static problems elipticity of the governing set of differential equations is no longer assured. As a consequence finite element analyses suffer from a severe mesh sensitivity. The deficiency of the standard continuum damage model can be circumvented by adding higher order terms in the constitutive law, which are thought to reflect phenomena at the microlevel, such as crack bridging phenomena and interactions among adjacent microcracks. Possible strategies are nonlocal and gradient damage models (Pijaudier-Cabot and Bazant 1987, Peerlings et al. 1995, de Borst et al 1996). These models introduce an internal length scale, which measures the distance over which a strong micro-structural interaction will exist. By the introduction of this internal length scale, the gradient damage model is capable of describing size effects (Carmeliet 1996).
An important issue is the proper experimental determination of the damage model parameters, such as the damage evolution law and the internal length scale. These parameters are not directly derivable from elementary tests. Recent developments of numerical analysis procedures in continuum mechanics coupled with optimization techniques permit the formulation of methods for performing inverse identification of the damage model parameters. These methods are formulated in the framework of system identification theory (Eykoff 1974, Beck and Arnold 1977). Aim of the system identification theory is not only to estimate the parameters, but also to analyze the identifiability problem, to evaluate the accuracy of the estimates and to formulate optimum experimental design criteria.

H. Sol and C.W.J. Oomens (eds.), Material Identification Using Mixed Numerical Experimental Methods, 203-212.

2. Gradient Damage Model

In this section, we present the gradient formulation of the nonlocal damage model as developed by Peerlings et al. (1995). The nonlocal damage model as well as the gradient damage model start from the classical isotropic elasticity-based damage theory:

$$\sigma = (1 - D)\, C\, \epsilon \tag{1}$$

where **C** is the initial elasticity tensor of the virgin material, σ and ϵ are the stress and strain tensors and D the damage variable, which grows from zero to one (complete loss of integrity). Damage growth is determined by an evolution law $D = F(\epsilon^{eq})$, in which ϵ^{eq} is an equivalent strain measure defined by Mazars (1984). Damage growth is possible if the damage loading function $f = \epsilon^{eq} - K$ vanishes. The damage parameter K initially equals the damage threshold K_0 and during damage evolution equals the maximum value of ϵ^{eq} ever reached during the loading history. The damage loading function f and the rate of damage growth $\dot{D}$ have to satisfy the discrete Kuhn-Tucker conditions.

In a nonlocal generalisation (Pijaudier-Cabot and Bazant 1987) the equivalent strain ϵ^{eq} is replaced by a spatially averaged or non-local equivalent strain value $\bar{\epsilon}$, such that:

$$\bar{\epsilon}(\mathbf{x}) = \frac{1}{V_r} \int_V \epsilon^{eq}(\mathbf{x}+\tau)\, \alpha(\tau)\, dV \quad , \quad \alpha(\tau) = \exp\left(-|\tau|^2 / 2 l^2\right) \tag{2}$$

with τ the separation vector, V_r a normalising factor, α a squared exponential weight function and l the so-called internal length scale.

In a gradient damage formulation of the nonlocal concept, the integral equation of eq. 2 is approximated by the partial differential equation (Peerlings et al. 1995):

$$\bar{\epsilon} - c\, \nabla^2 \bar{\epsilon} = \epsilon^{eq} \tag{3}$$

with $c = l^2/4$. The equilibrium equation and differential equation (3) are solved by means of the finite element method.

3. Markov Estimation

Let **Y** be a vector containing the response variables of a structure, such as forces, displacements. The vector **b** contains the model parameters to be estimated and **X** is the vector denoting the basic variables independent on the parameters, e.g. the geometry and boundary conditions. Consider a model description in the form $\eta = f(\mathbf{X}, \mathbf{b})$, with η the model response vector, which in many cases is only available by means of a nonlinear finite element algorithm. The differences between measured responses, $\mathbf{Y}_m$, and the model responses, η, are defined as the error $\mathbf{e} = \mathbf{Y}_m - \eta$. The basic estimation problem includes the development of knowledge about the parameters **b** based on the observation $\mathbf{Y}_m$. This knowledge can be expressed in terms of a probability density function $p[\beta]$, where β represents the estimate of the parameter vector **b**. The complete determination of the probability density function is in most cases quite complicated and impractical. The interest is generally reduced to the most significant characteristics of $p[\beta]$: the expected value $E[\beta]$ and the covariance $\mathrm{cov}[\beta] = E[(\beta - E[\beta])(\beta - E[\beta])^T]$.

In addition to the measurement observation a priori information on the measurement and parameters in the form of probability density functions may be available. In the literature (Eykoff 1974, Beck and Arnold 1977) a number of different estimation procedures have been developed, which mainly differ in the use of a priori information. For the *least-square* method, the only assumption is that the measurements can be successfully approximated by the model. The *Markov* estimator requires as well the knowledge of the error structure by means of the measurement error covariance $\mathbf{C}_e = \mathrm{cov}[\mathbf{e}]$. For the *maximum likelihood* estimator, one needs to know the probability density of the measurement, while the *Bayes'* estimation procedure requires as well knowledge of the a priori probability density function of the unknown parameters.
Since the a priori information is usually difficult to obtain, we will use the Markov method for the estimation of the gradient damage parameters. If the measurements are independent and the errors have zero mean, $E[e]=0$, the measurement error covariance matrix reduces to a diagonal matrix, where the diagonal elements represent the error variances s_e^2. To get information on the measurement error variances, the standard deviation of repeated experiments under the same conditions may be used.
The Markov estimation method implies the minimization of an objective function $S=\mathbf{e}^T\mathbf{C}_e^{-1}\mathbf{e}$. A necessary condition to minimize S is that the matrix derivative of S with respect to $\mathbf{b}$ be equal to zero. The matrix derivative is given by:

$$\frac{\partial S}{\partial \mathbf{b}} = -2\,\mathbf{X}^T\mathbf{C}_e^{-1}\,\mathbf{e} \quad , \quad \mathbf{X} = \frac{\partial \eta}{\partial \mathbf{b}} \tag{4}$$

where $\mathbf{X}$ represents the sensitivity matrix. The Markov estimator can be proved to be unbiased, $E[\boldsymbol{\beta}]=\mathbf{b}$, and to have the property of minimum variance for all unbiased estimates. For a Markov estimator the parameter covariance matrix $\mathrm{cov}[\boldsymbol{\beta}]$ at the optimal point is given by:

$$\mathbf{P} = \mathrm{cov}[\boldsymbol{\beta}] = \left(\mathbf{X}^T\mathbf{C}_e^{-1}\mathbf{X}\right)^{-1} \tag{5}$$

The diagonal elements of the matrix $\mathbf{P}$ represent the variances of the parameter estimates s_β^2, which are a measure for the accuracy of the estimations. The ijth element of the correlation matrix $\mathbf{R}$ is given by:

$$\mathbf{R}_{ij} = \frac{\mathbf{P}_{ij}}{\sqrt{\mathbf{P}_{ii}\,\mathbf{P}_{jj}}} \tag{6}$$

The off-diagonals of $\mathbf{R}$ provide the correlation coefficients between the different parameters. The values of the off-diagonals range between -1 and 1. To solve the unconstrained minimization of the objective function S the Gauss-Newton method and the Levenberg-Marquadt algorithm can be used. Starting from a set of parameter estimates $\boldsymbol{\beta}$, a correction is found by means of:

$$\Delta\boldsymbol{\beta} = \left(\mathbf{X}^T\,\mathbf{C}_e^{-1}\mathbf{X} + \lambda\,\Omega\right)^{-1}\mathbf{X}^T\,\mathbf{C}_e^{-1}\,\mathbf{e} \tag{7}$$

where $\lambda\Omega$ is a matrix to reduce the size and direction of the step. For $\lambda=0$ the Gauss-Newton method is found, for $\Omega=I$ and very large λ the method of steepest descent is recovered.

4. Experimental Study

The material studied is a cellulose fibre reinforced cement mortar. The identification experiments consist of uniaxial tensile tests on double notched specimens and notched four point bending tests of different sizes. The test configurations are given in figure 1. All the specimens have a thickness of t=7.3 mm. The dimensions of the tension specimen are: L=56 mm, B=28 mm and a_0=5 mm. The measurement length for the displacement u is L_m=50 mm.

The size range of the geometrically similar bending specimens is 1 to 8: H=24 mm (S1), H=48 mm (S2), H=96 mm (S4), H=192 mm (S8). The beam span L=35/8H and the distance between the external forces W=L/3. Two notch lengths are considered: a_0=H/6 (N1) and a_0=H/3 (N3). The crack-mouth opening displacement v(1) is measured over the length L_m=10 mm. The tests are repeated for similar samples. Figure 2 shows as example the measured load-cmod curves for the different beam sizes. The nominal stress for the tensile test is defined as $\sigma_t = F/(B-2a_0)t$, with F the load. The nominal stress for the bending specimen is defined as $\sigma_b = cFW/(H-a_0)^2t$. To make σ_b for convenience coincide with the maximal tensile stress according to the theory of bending, the constant c is taken equal to 1.5. The average nominal strength $\mu_{\sigma m}$ and displacement μ_{um} at maximum loading together with the corresponding standard deviation $s_{\sigma m}$ and s_{um} for the different specimens are listed in table 1.

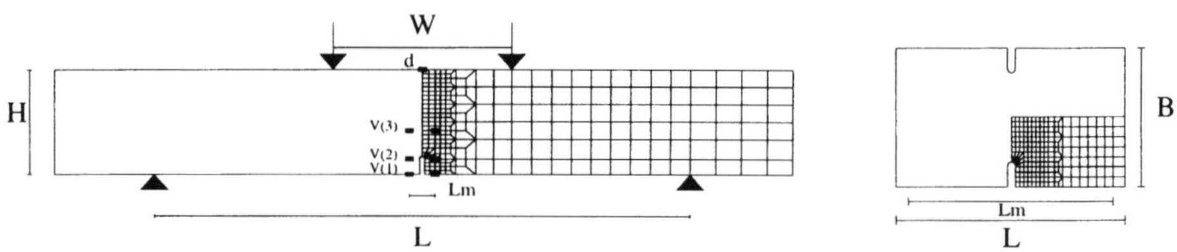

Figure 1. Geometry and finite element representation of bending and four point bending test

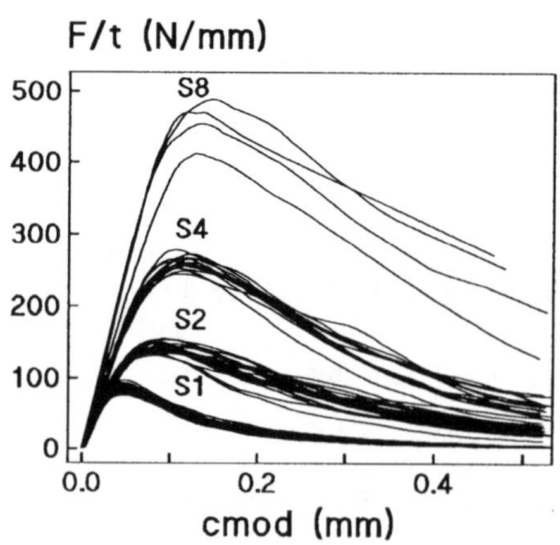

Figure 2. Load-cmod curves for four bending tests of different sizes

TABLE 1. Testing results. Average nominal strength $\mu_{\sigma m}$ and corresponding displacement μ_{um} for the double notched tensile and four point bending tests. The accuracy is specified by means of the standard deviations s_{um}, based on repetitions of the experiment.

experiment type	number	$\mu_{\sigma m}$	$s_{\sigma m}$	μ_{um}	s_{um}
T	51	12.33	1.28		
BS(1)N(1)	36	22.5	1.37	0.046	0.0029
BS(2)N(1)	24	18.6	0.78	0.088	0.0068
BS(3)N(1)	14	17.1	0.58	0.117	0.066
BS(4)N(1)	4	15.0	-	0.132	-
BS(1)N(3)	46	24.2	1.09	0.068	0.0061
BS(2)N(3)	30	21.0	0.94	0.134	0.0132
BS(3)N(3)	8	18.9	0.98	0.166	0.0097
BS(4)N(3)	2	15.2	-	0.203	-

Codes : T=tensile, B=bending, S=size, N=notch, v=crack-opening displacement, u=displacement, d=deflection, σ_{tm}=nominal tensile strength, σ_{bm}=nominal bending strength.

5. Gradient Damage Parameter Estimation

We assume that the material is homogeneous and can be modelled by the isotropic gradient damage model. For the damage function we adopt an exponential function:

$$D = 1 - (K_0/\bar{\epsilon})[(1 - a_D) + a_D \exp(-b_D(\bar{\epsilon} - K_0))] \tag{8}$$

with a_D=0.95. The model parameters to be identified are the Young's modulus E, the tensile strength f_t, the damage evolution parameter b_D of eq. 8, and the internal length scale l. The Young's modulus E is, after correcting for the presence of the notches, directly obtained from uniaxial tensile tests. The tensile strength f_t, the damage evolution parameter b_D and the internal length scale l are identified using the Markov estimation method.

To initiate the iterative estimation process, an initial guess of the parameters β_0 is needed. The nominal tensile strength σ_{tm} yields a reasonable initial estimate for the tensile strength f_t. To initially estimate the internal length l and damage evolution parameter b_D, the following procedure is followed. First initial guesses of the parameters are obtained by trial and error fitting of the nominal strength σ_{bm} for the smallest four point bending beam (S1). It is found that various sets of the parameters fit the nominal strength σ_{bm}: l=1mm b_D=15.5, l=2mm b_D=42, l=4mm b_D=296. Next, the parameter set which fits the measured size effect the best is selected as initial estimate (figure 3). For the Markov estimation of the parameters we will use the nominal tensile strength σ_{tm} and the 8 nominal bending strengths σ_{bm} as listed in table 1. The standard deviations $s_{\sigma m}$ give the necessary information on the measurement error variances s_e^2, which represent the diagonal elements of measurement error covariance matrix C_e. The sensitivity coefficients are calculated using a central finite difference approximation. The estimation is repeated over until all parameters have reached

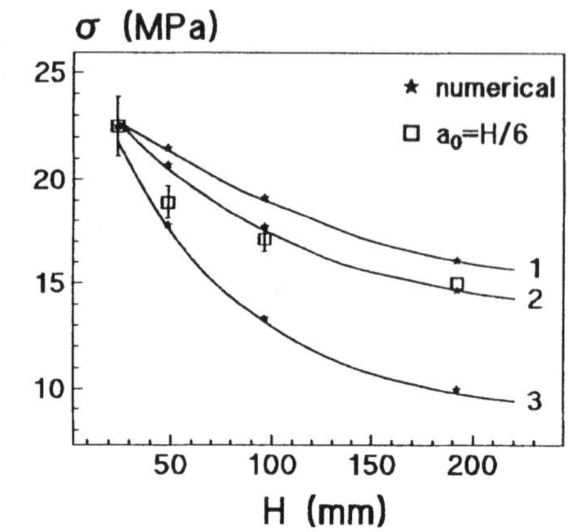

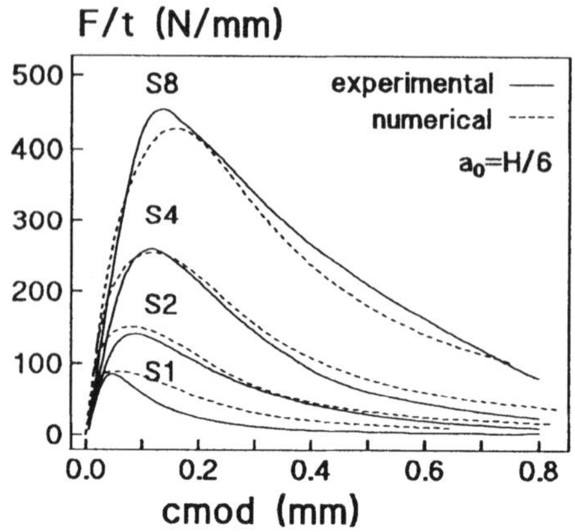

Figure 3. Comparison of the measured and calculated size effect for sets of internal lengths scales l and damage evolution parameters b_D: (1) l=1mm b_D=15.5, (2) l=2mm b_D=42, (3) l=4mm b_D=296

Figure 4. Comparison of the calculated and measured nominal stress-cmod curves for different beam sizes (S1-2-4-8)

convergence. The estimation results $\boldsymbol{\beta}$ are listed in table 2. To validate the obtained parameters, the numerical and experimental load-cmod curves are compared in figure 4 for different beam sizes. Table 3 gives the obtained parameter variances and correlation coefficients. It can be observed that the cross-correlation are all lower than 0.9, which means that the nine measured nominal strengths contain sufficient information to independently estimate the three unknown parameters. On the other hand, we notice the relative large standard deviation for the internal length scale and damage parameter. These high values may be caused by the relative large values of the measurement error variances or indicate that a poor experimental design is used to estimate the model parameters.

6. Optimal Design of experiments

This section deals with the selection of measurement information to maximize the accuracy of the estimated model parameters. A common measure of the accuracy of an estimator is its variance: the smaller the variance, the greater the precision. More general criteria have been suggested based on the minimization of the confidence interval around the estimated parameters. For additive, zero mean and normal errors the minimization of the confidence region is equivalent to the minimization of the determinant of the parameter covariance matrix $|\mathbf{P}|$, or equivalently the maximization of the determinant $|\mathbf{X}^T\mathbf{C}_e^{-1}\mathbf{X}|$. Another issue related to the design of optimal experiments is the identifiability problem. This problem investigates the conditions under which it is not possible to independently estimate all the parameters from the selected measurements. It can be shown that the parameters cannot all be uniquely

estimated if the sensitivity coefficients in the neighbourhood of the minimum of the sum of squares are linearly dependent. This criterion applies for linear and nonlinear estimation problems. If the second order derivatives $\beta_i\beta_j\partial^2\eta/\partial\beta_i\partial\beta_j$ are negligible, this criterion is equal to the determinant $|\mathbf{X}^T\mathbf{C}_e^{-1}\mathbf{X}|$ not to be zero.
The minimization of the variances and the non-zeroness of the determinant $|\mathbf{X}^T\mathbf{C}_e^{-1}\mathbf{X}|$ will be used as criteria to evaluate the accuracy of the minimum variance estimators.
We will analyze the accuracy and identifiability of the parameter estimation problem for different combination of test setups and location of response sensors. Two different methods will be considered. The same material data as determined in the parameter estimation problem are used (table 2). The geometry of the tensile and four point bending specimen are given in figure 1.

6.1 METHOD I. USE OF THE COMPLETE LOAD-DISPLACEMENT CURVES

In a first method the measurement information consists of complete nominal stress-displacement curves of the tensile (T) and four point bending tests (B), including the descending branch. To perform the analysis information on the measurement error variances as function of the loading history are needed. We consider the displacements, i.e. the elongation and the cmod for the tensile and four point bending test respectively, as the independent variables controlling the loading history of the specimen. Then, the measurement error variance can be calculated by means of the standard variation of the nominal stress s_σ at uniformly spaced displacements in the total displacement range. The standard deviations s_σ were found to be approximately proportional to the nominal stress σ. For the tensile specimen a constant coefficient of variation was found: $s_\sigma/\sigma_t \approx 0.11$. For the four point bending experiments the coefficient of variation s_σ/σ_b of the nominal stress as function of the dimensionless displacement u/u_m, where u_m is the displacement at maximum loading, is plotted in figure 5. We observe that the curves for different sizes approximately coincide. They can be described by the empirical model $s_\sigma/\sigma = 0.043u/u_m + 0.2\exp(-4.6u/u_m)$. The sensitivity coefficients are obtained by using central finite differences. The scaled sensitivity coefficients $Z_i = \beta_i X_i/s_i$ for the three parameters are plotted in figure 6.
In table 4 the results are given for various combinations of measurement information: (2a) the tensile stress-displacement curve only; (2b) the stress-cmod curve of the bending specimen only; (2c) the stress-displacement curve of the tensile and bending specimen. We begin with the results for the tensile test (case 2a). All the cross-correlations exceed 0.99, which means that the variances are highly correlated and inaccurate. The nearly zero value of the determinant $|P|$ indicates that the reason for these high correlations among the parameters is the nearly linear dependence of the sensitivity coefficients. The nearly linear dependence especially in the post peak range becomes also clear from figure 6. Due to the linear dependence of the sensitivity coefficients the objective function has no unique minimum point and the model parameters cannot estimated without ambiguity. For the cases 2b and 2c the problem of identifiability is solved. However, since high dependencies among the parameters still exist the parameters cannot independently determined. Therefore, the estimation of the

model parameters from complete stress-displacements curves of one specimen size cannot be recommended as identification method.

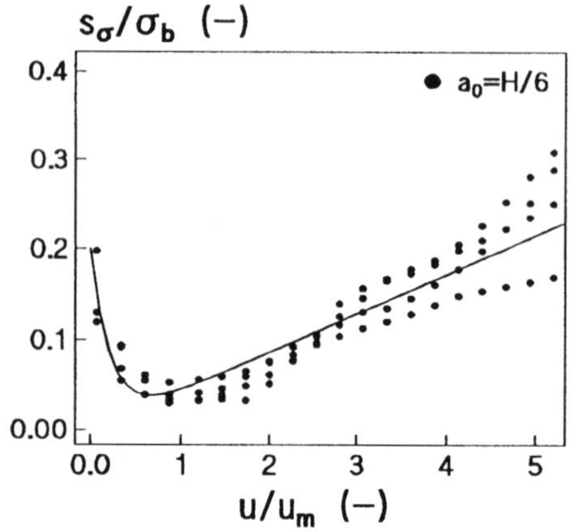

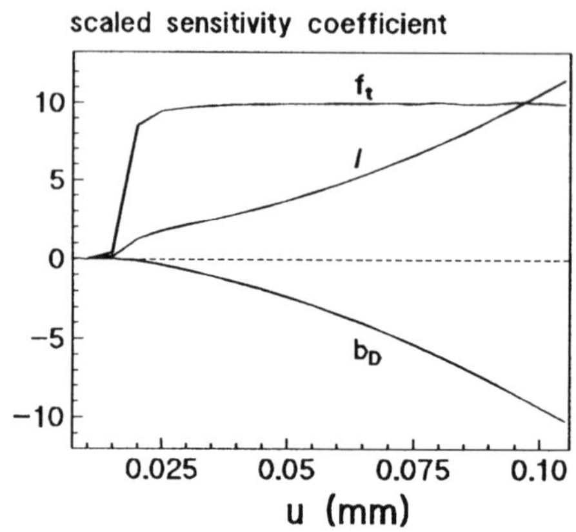

Figure 5. Four point bending test of beam size 2. Evolution of the coefficient of variation s_σ/σ_b in function of the dimensionless cmod u/u_m, where u_m is the cmod at maximum loading (• measurements, — model).

Figure 6. Evolution of the scaled sensitivity coefficients $Z_i=\beta_i X_i/s_i$ in function of the displacement: four point bending size S2.

6.2 METHOD III DIFFERENT DISPLACEMENT SENSORS

In this method the measurement data consist of the nominal tensile strength σ_{tm}, the nominal strength of one-size four point bending beam σ_{bm} and additional information by means of local displacements. The considered displacements at maximum loading are: the deflection d, the crack-mouth opening displacement v(1), the crack opening displacement at the notch tip v(2), the crack opening displacement at a fixed position along the crack trajectory v(3). The measured standard deviations $s_{\sigma m}$ and s_{um} of table 1 provide the necessary information for the diagonal elements of the measurement error covariance C_e. Table 5 lists the estimation results in order of decreasing additional information on the displacements. It is clear that for all the cases the parameters can be independently estimated, since the values of the correlation coefficients are lower than 0.9. Note that the standard deviations obtained with this method are smaller than for the previous methods. Comparison of the different cases indicates that the deflection does not give relevant additional information when crack opening displacements are already available. On the other hand, detailed information on the displacement along the notch and crack trajectory increases the accuracy of the estimated parameters.

TABLE 2. Parameter estimation results using the nominal tensile strength σ_{tm} and 8 nominal bending strengths σ_{bm}. Expected values of the parameters.

l (mm)	b_D (-)	f_t (MPa)	E (MPa)
2.4	42.6	11.18	12860 (2837*)

TABLE 3. Parameter estimation results using the nominal tensile strength σ_{tm} and 8 nominal bending strengths σ_{bm}.

type	specimen code	Sensor code	s_l mm	s_{bD} -	s_{ft} MPa	$\rho_{l,bD}$ -	$\rho_{l,ft}$ -	$\rho_{bD,ft}$ -
1	BS(1248)N(1), BS(1248)N(3), T	σ_{bm} σ_{bm} σ_{tm}	0.81	10.5	1.4	0.02	0.88	0.47

TABLE 4. Parameter estimation results using complete nominal stress-displacement curves of the tensile test and/or bending test of size S2.

type	specimen code	Sensor code	s_l mm	s_{bD} -	s_{ft} MPa	$\rho_{l,bD}$ -	$\rho_{l,ft}$ -	$\rho_{bD,ft}$ -
2a	T	σ_t-u	19.9	207	7.2	0.99	0.99	0.99
2b	BS(2)N(1)	σ_b-v(1)	1.6	22.0	1.12	0.99	0.94	0.88
2c	T BS(2)N(1)	σ_t-u σ_b-v(1)	0.63	10.7	0.43	0.98	0.80	0.71

TABLE 5. Parameter estimation results using the nominal strength σ_{tm}, σ_{bm} and displacement information for one-size specimens.

type	specimen code	Sensor code	s_l mm	s_{bD} -	s_{ft} MPa	$\rho_{l,bD}$ -	$\rho_{l,ft}$ -	$\rho_{bD,ft}$ -
3a	BS(1)N(1) T	σ_{bm},v(123),d σ_{tm}	0.53	6.06	0.73	0.58	0.78	0.01
3b	BS(1)N(1) T	σ_{bm},v(123) σ_{tm}	0.55	6.09	0.75	0.58	0.8	0.01
3c	BS(1)N(1) T	σ_{bm},v(12) σ_{tm}	0.84	11.2	1.15	028	0.8	0.31
3d	BS(1)N(1) T	σ_{bm},v(1) σ_{tm}	0.9	13.5	1.16	0.4	0.76	0.26

Codes : T=tensile, B=bending, S=size, N=notch, v=crack-opening displacement, u=displacement, d=deflection, σ_{tm}=nominal tensile strength, σ_{bm}=nominal bending strength.

7. Conclusions

1. A general formulation to perform inverse identification of the gradient damage model parameters based on a Markov estimator has been presented. The method provides information on the accuracy and uniqueness of the estimated values. The approach also offers possibilities for the design optimization of laboratory tests in order to minimize the effect of errors on the estimated values.
2. The gradient damage parameters cannot be accurately estimated from load-displacement curves of one-size tensile and four point bending tests. If additional information in the form of local deformations along the crack trajectory is supplied, accurate estimation can be obtained.
3. The presented estimation method does not consider modelling errors, but is based on the assumption that the model is correct. The estimation procedure may be modified in order to investigate the validity of different competing models. The method of model discrimination involves the design of experiments, which maximize the difference between the predicted responses of two or more models in order to select the most suitable model.
4. In this analysis the material was assumed to be homogeneous. However, the heterogeneity of quasi-brittle materials may play an important role in the localization process and damage behaviour of structures. Therefore, the use of stochastic approaches to the damage process seems essential (Carmeliet and Hens 1994, Carmeliet and de Borst 1995). Then, the model parameters are random correlated variables and adequate estimation methods should be used.

8. References

Carmeliet, J. and Hens, H. (1994). Stochastic nonlocal continuum damage model for continua with random field properties. *ASCE J. Engng. Mech.* **120**, 2013-2027.

Carmeliet, J., and de Borst, R. (1995). Stochastic approaches for damage evolution in standard and non-standard continua. *Int. J. Solid Structures.* **32**, 1149-1160.

Carmeliet, J. (1996) Gradient damage and size effects. in Y.K. Lin and T.C. Su (eds), *Proc. of the 11th Conference on Engineering Mechanics*, ASCE, New York, pp. 1175-1178,.

de Borst R, Pamin J. Peerlings R.H. and Sluys L.J. (1995). On gradient-enhanced damage and plasticity models for failure in quasi-brittle and frictional materials, *Comput. Mech.* **17**, 130-141.

Eykoff P. (1974). *System identification. Parameter and State Estimation.* John Wiley and sons, Chisester.

Mazars, J. (1984). Application de la mécanique de l'endommagement au comportement non linéaire et à la rupture du béton de structure. These d'État, Université Paris VI, Paris.

Peerlings, R.H.J., de Borst, R., Brekelmans, W.A.M., and de Vree,J.H.P. (1995) Computational modelling of gradient-enhanced damage for fracture and fatigue problems. *Proc., Computational Plasticity: Theory and Applications* (Edited by D.R.J. Owen *et al.*), pp. 75-986 , Pineridge Press, Swansea.

Pijaudier-Cabot, G. and Bažant, Z.P. (1987). Nonlocal damage theory. *ASCE J. Engng. Mech.* **113**, 1512-1533.

PARAMETER IDENTIFICATION OF THE COHESIVE CRACK MODEL

G. BOLZON, D. GHILOTTI and G. MAIER
Department of Structural Engineering
Politecnico (Technical University) of Milan
piazza Leonardo da Vinci 32, 20133 Milan, Italy

Abstract — This paper concerns the inverse problem of identifying the parameters of the "cohesive crack model" for quasi-brittle fracture simulations formulated as linear complementarity problems. These parameters are indirectly determined by measuring through optical means (laser interferometry) the displacements of points on the specimen surface during three-point-bending tests. The identification procedure, resting on a deterministic least-square approach, amounts to a constrained minimisation of a quadratic function measuring the discrepancy between computed and measured displacements. The nonconvex nonlinear programming problem thus arrived at is numerically solved by different techniques, whose computational effectiveness is comparatively discussed.

1. Introduction

Fracture processes in concrete, rocks, ceramics, mortars, bone tissues and other quasi-brittle materials, are often numerically simulated, to structural analysis purposes, on the basis of the "cohesive crack model" (Hillerborg, 1991; Planas and Elices, 1991; Guinea, 1995). This model is centred on the assumption that the only nonlinearity consists of a softening interface law along a locus Γ_p of possible displacement discontinuities, see Figure 1a, while elsewhere in the domain Ω the material behaviour is linear elastic.

For concrete-like materials and mode I fracture (which prevails in crack propagation processes), the interface constitutive law in terms of normal traction p versus normal displacement jump w, was found to be realistically interpretable as piecewise-linear (PWL), see Figure 1b, on the basis of extensive laboratory experiments (e.g. Wittmann and Hu, 1991). Namely, the decay to zero of p as w grows is governed by four material parameters, say: the tensile strength p_c; the coordinate p_b of the "breakpoint" in the sloping down p versus w plot; the initial slope k; the parameter h which relates k to the slope h^* of the second descending linear branch, $h^* = hk/(h+k)$.

This paper concerns the inverse problem of identifying the above material parameters on the basis of standard three-point-bending (3PB) tests. This experimental technique is widely used for the characterisation of quasi-brittle materials.

H. Sol and C.W.J. Oomens (eds.), Material Identification Using Mixed Numerical Experimental Methods, 213-222.

For concrete and mortar, the 3PB test is often performed according to RILEM (1985) rules in order to determine the fracture energy G_f which represents the area under the *p* versus *w* curve (Hillerborg, 1985). Various proposals are available in the literature on the use of 3PB tests for the identification of the above defined four parameters (Figure 1b), see e. g. Guinea *et al.* (1994) and the papers quoted therein.

Most approaches to determine the fracture properties of quasi-brittle materials use information from the load-displacement curve (i.e. Q versus q with the symbols of Figure 1a), see e.g. the recent work by Goretti *et al.* (1995) and by Nanakorn and Horii (1996).

To parameter identification purposes, the measurements performed in 3PB tests may concern either (i) the imposed displacement and the relevant reactive force all along the experiment, or (ii) displacements of numerous points, captured by optical means (e.g. by laser interferometry), in a few static situations suitably selected during the test. These two kinds of measurement techniques naturally suggest different identification strategies: (i) methods, like Kalman-Bucy filters (cp. Bittanti *et al.*, 1983) apt to exploit at best the flow of experimental information on few quantities (also in a non-deterministic context); (ii) minimisation of a suitable discrepancy function enforcing (e.g. by constraints) all relations governing the underlying analysis problem, in a basically deterministic context.

In the inverse procedure considered here, the main experimental data are of the above kind (ii), namely they consist of the displacements of some reference points on the surface of the specimen, in a grid Gr ≡ {P1, P2, ..., Pn}, as shown in Figure 1a. The mathematical model used for the simulation of the test has the format of a linear complementarity problem (LCP), i.e. a linear equation in two variable vectors which are sign-constrained and required to be orthogonal. This particular mathematical structure is generated by the holonomic (i. e. reversible, path-independent) description of the aforementioned local interface PWL law, and is carried over to the overall analysis model through spatial discretisation by finite or boundary elements, as shown by Bolzon *et al.* (1994). The identification problem is then formulated, by a deterministic least-square approach, as the minimisation of a quadratic discrepancy function subjected to nonconvex inequality constraints.

Earlier theoretical and numerical work on identification related to LCP

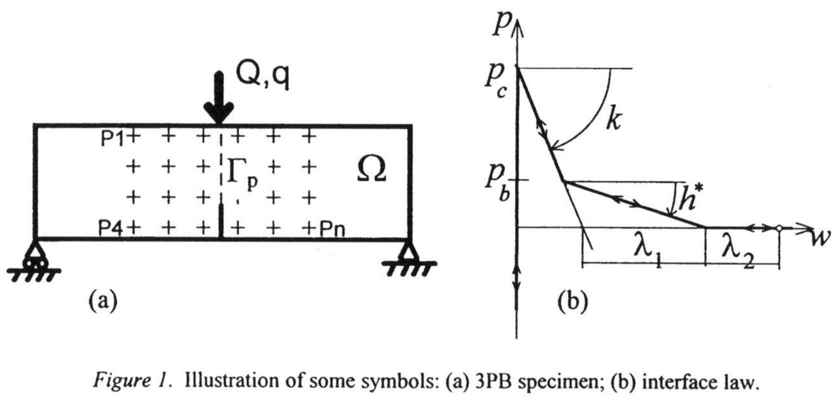

Figure 1. Illustration of some symbols: (a) 3PB specimen; (b) interface law.

analysis (Maier, 1981; Maier *et al.*, 1982) was characterised by the fact that the parameters to identify were linearly contained in the LCP vector, while herein they are included in the LCP matrix and, hence, induce a nonlinearity in the minimisation constraints.

2. The cohesive crack model

In 3PB tests, crack propagation can a priori be conjectured to occur along the symmetry axis, which is therefore considered as the locus Γ_p of possible discontinuity of displacements, or briefly "interface". Since the external action (the load factor or, more frequently, the amplitude of a controlled displacement) is monotonically increased in time, local unloading is unlikely to occur. The interface law can then be conceived as holonomic (i. e. path-independent, reversible, nonlinear elastic). With such assumption, the analytical description of the local behaviour depicted in Figure 1b, can be cast into the following LCP format (Bolzon *et al.*, 1994):

$$\varphi = p_c \mathbf{v}_1 - p_b \mathbf{v}_2 - [k\,\mathbf{m}_1 + h\mathbf{m}_2]\lambda - p\mathbf{n} \geq \mathbf{0}, \quad \lambda \geq \mathbf{0}, \quad \varphi^{\mathrm{T}}\lambda = 0 \tag{1}$$

where $\varphi^{\mathrm{T}} \equiv \{\varphi_1\ \varphi_2\ \varphi_3\}$, $\lambda^{\mathrm{T}} \equiv \{\lambda_1\ \lambda_2\ w\}$ are the vectors which collect the main variables, while $\mathbf{n}$, $\mathbf{v}_1$, $\mathbf{v}_2$, $\mathbf{m}_1$, $\mathbf{m}_2$ are vectors and matrices of constant entries:

$$\mathbf{n} = \begin{Bmatrix} 0 \\ 0 \\ 1 \end{Bmatrix}, \quad \mathbf{v}_1 = \begin{Bmatrix} 1 \\ 1 \\ 1 \end{Bmatrix}, \quad \mathbf{v}_2 = \begin{Bmatrix} 1 \\ 0 \\ 0 \end{Bmatrix}, \quad \mathbf{m}_1 = \begin{bmatrix} -1 & -1 & 1 \\ -1 & -1 & 1 \\ -1 & -1 & 1 \end{bmatrix}, \quad \mathbf{m}_2 = \begin{bmatrix} -1 & 0 & 0 \\ 0 & 0 & 0 \\ 0 & 0 & 0 \end{bmatrix} \tag{2}$$

The frequently adopted single linear softening branch (used later within the proposed identification procedure) can be obtained as a special case of (1):

$$\varphi = p_c \mathbf{v}_1 - k\,\mathbf{m}_1 \lambda - p\mathbf{n} \geq \mathbf{0}, \quad \lambda \geq \mathbf{0}, \quad \varphi^{\mathrm{T}}\lambda = 0 \tag{3}$$

In this case, $\varphi^{\mathrm{T}} \equiv \{\varphi_1\ \varphi_2\}$, $\lambda^{\mathrm{T}} \equiv \{\lambda\ w\}$, and:

$$\mathbf{n} = \begin{Bmatrix} 0 \\ 1 \end{Bmatrix}, \quad \mathbf{v}_1 = \begin{Bmatrix} 1 \\ 1 \end{Bmatrix}, \quad \mathbf{m}_1 = \begin{bmatrix} -1 & 1 \\ -1 & 1 \end{bmatrix}. \tag{4}$$

3. The mathematical model by simulations of 3PB tests

Let Ω be the two dimensional domain occupied by the 3PB specimen under consideration interpreted as a plain-stress system. According to the cohesive crack model, all non-linearities are supposed to be concentrated along the interface Γ_p (i. e. in the locus of a mode I fracture process, on the symmetry axis). Since linear elasticity is assumed elsewhere in Ω the superposition can be used to relate the distribution of traction $p(x)$ to the displacement discontinuity $w(x)$:

$$p(x) = p^E(x) + \int_{\Gamma_p} Z(x,x')\, w(x')\, d\Gamma \qquad x,x' \in \Gamma_p \tag{5}$$

where: $p^E(x)$ is the traction generated on Γ_p by the external action, consisting here of the controlled displacement q, in the absence of displacement discontinuities; $Z(x,x')$ denotes the Green (influence) function which reflects the geometric and elastic properties of the considered specimen and represents the effect of the displacement discontinuity in the otherwise unloaded body.

The overall mathematical model for the simulation of the 3PB test turns out to be a linear complementarity problem (LCP) resulting from the combination of Equation (5) with the above PWL constitutive laws, Equations (1) or (3).

To numerically solve the analysis problem, space discretisation has to be performed either by the finite element method (Bolzon, 1996) or by a symmetric Galerkin boundary element method (Maier *et al.*, 1993). The unknown fields p and w along Γ_p are also modelled by means of suitable interpolation functions, so that these fields are governed by the vectors **P** and **W** of conjugate generalised variables in Prager's sense. Thus, the discretised version of Equation (5) results into a system of linear equations: $\mathbf{P} = \mathbf{P}^E + \mathbf{Z}\,\mathbf{W}$, where **Z** is a symmetric negative-semidefinite matrix of influence coefficients. The interface law is also re-formulated in a weighted averaged sense, leading to a LCP in terms of the above generalised variables **P** and **W** and in terms of the conjugate generalised variables Λ and Φ, which are the discrete counterparts of the local variables λ and φ, see Maier *et al.* (1993), Bolzon *et al.* (1994).

Combination of the above discretised linear equilibrium equations with the nonlinear interface law, gives the final version of the direct (analysis) problem in the LCP format:

$$\Phi = p_c \mathbf{V}_1 - p_b \mathbf{V}_2 - [k\mathbf{M}_1 + h\mathbf{M}_2 + \mathbf{Z}']\Lambda - \mathbf{P}^E \geq \mathbf{0}, \;\; \Lambda \geq \mathbf{0}, \;\; \Phi^T \Lambda = 0 \tag{6}$$

where $\mathbf{V}_1$, $\mathbf{V}_2$, $\mathbf{M}_1$ and $\mathbf{M}_2$ are the discretised counterparts of vectors and matrices defined by (2), while $\mathbf{Z}' = diag[\mathbf{0} \;\; \mathbf{0} \;\; \mathbf{Z}]$ when $\Lambda^T = \{\Lambda_1^T \;\; \Lambda_2^T \;\; \mathbf{W}^T\}$.

Note that the dimension of the LCP (6) is equal to $3n$ ($2n$ for a single softening branch), n being the number of the discretisation nodes on Γ_p.

4. The parameter identification problem

Consistently with the assumptions of Section 3, let us evaluate the displacements $\mathbf{U}^c$, in a chosen direction, of the points belonging to the grid $Gr \equiv \{P1, P2, ..., Pn\}$ and the reaction Q^c. To this aim superposition effects in the hypothetical undamaged body can be expressed in generalized variables as follows:

$$\mathbf{U}^{c} = \mathbf{U}^{E} + \mathbf{G}\,\mathbf{W}; \qquad Q^{c} = Q^{E} + \mathbf{R}\,\mathbf{W} \tag{7}$$

where $\mathbf{U}^{E}$ and Q^{E} denote elastic responses to the external action only, **G** and **R** are the effects of the displacement discontinuities **W**.

The experimental data corresponding to $\mathbf{U}^{c}$ and Q^{c} are indicated by $\mathbf{U}^{m}$ and Q^{m}. The components of vector $\mathbf{U}^{m}$ may be measured by the electronic speckle pattern interferometry (ESPI) technique, see e. g. Facchini and Zanetta (1996).

The set of the present unknown fracture parameters (p_c, p_b, k and h) can be evaluated by minimising the discrepancy (objective) function defined by

$$\omega \equiv \left(\mathbf{U}^{m} - \mathbf{U}^{c}\right)^{T} \mathbf{D}_{U} \left(\mathbf{U}^{m} - \mathbf{U}^{c}\right) + D_{Q}\left(Q^{m} - Q^{c}\right)^{2} \tag{8}$$

Here $\mathbf{D}_U$ and D_Q represent a positive definite symmetric weighting matrix and a positive weighting factor, respectively. In this paper, no statistical characterisations of uncertainties (e.g. of experimental errors) have been allowed for. Matrix $\mathbf{D}_U$ has been assumed diagonal, with non-zero entries equal to $\left(U^{m}_{max}\right)^{-2}$, and D_Q has been set equal to $\left(Q^{m}\right)^{-2}$ in order to render ω non-dimensional.

The quantities $\mathbf{U}^{c}$ and Q^{c} in the objective function (8) depend on the material parameters through Equations (7) supplemented by Equations (6). The inverse problem thus formulated can be re-written as :

$$\min_{\substack{p_c, p_b, k, h \\ \Lambda_1, \Lambda_2, \mathbf{W}}} \left\{ \omega \equiv \mathbf{W}^{T}\mathbf{C}\,\mathbf{W} + \mathbf{d}^{T}\mathbf{W} + e \right\} \tag{9}$$

subjected to the constraints given by the inequalities and the orthogonality condition of relations (6). Matrix **C**, vector **d** and the constant e follow in a straightforward way from Equations (7) and (8).

Upper and lower bounds on the values of the material parameters to be found (p_c, p_b, k and h), are introduced on the basis of a piori "expert knowledge" or engineering judgement. This generally entails a faster convergence toward the solution of the inverse problem in point.

5. Minimisation procedures

In order to find the minima of the convex function (9) subjected to the nonlinear nonconvex constraints (6), two different algorithms have been employed, among the many proposed in the literature (see e.g. Arora *et al.*, 1995), namely: a sequential quadratic programming (SQP) method and a genetic algorithm (GA).

The SQP algorithm coded in Matlab (1996) package is designed to numerically solve general constrained optimisation problems, with recourse to the Lagrangian which incorporates the objective function and the original constraints. A sequence of convex quadratic approximations to this Lagrangian is hence built up by using information from the gradients only (no second derivatives are evaluated). At each iteration the global optimum point is computed (by linear search) recursively along the direction which minimises the convex quadratic approximations constructed in the current iteration.

In the computational tests run so far to minimise the objective function ω (9) subjected to the constraints (6), convergence was not easily achieved unless the complementarity condition (6c), which is known to be a severe restriction to the use of many optimisation procedures (Luo *et al.*, 1996), has been relaxed. A good convergence was instead observed after replacing the original problem by its penalised version, i. e. by adding to the objective function the term $\chi\ \Phi^T\Lambda$, where $\chi>0$ is the penalty factor, to be chosen on the basis of preliminary numerical tests.

Genetic algorithms have been shown to be effective in numerous kinds of optimisation problems (Goldberg, 1989; Davis, 1991).

In the present application, the genetic procedure is started with a randomly generated population, each member of which represents a set of the parameters p_c, p_b, k and h, inside the 4-dimensional box defined by upper and lower bounds which are chosen on the basis of a priori conjectures.

For each member (individual) of the population, the/a solution to the LCP (6) is computed in terms of variables Λ (and hence $\mathbf{W}$) and Φ, using the techniques discussed in Bolzon *et al.* (1995). From these solutions, the displacements $\mathbf{U}^c$ and the reaction Q^c to be entered in the objective function (8), can be easily recovered through equations (7). Note that for some sets of the four material parameters, characterizing very brittle material behaviour, the LCP (6) may admit multiple solutions under displacement control (here assumed) as a consequence of snap-back instability, see e.g. curves (b) and (c) of Figure 2.

The fitness of each individual (and of each corresponding LCP solution) is assessed on the basis of the corresponding value of the objective function. A subset of better fit individuals is then selected from the initial population and from each one of the subsequent generations, and is used to reproduce a new population by operations like cross-over and mutation. The process of reproduction is continued until further reduction of the objective function is no longer observed in the fittest member of the population.

As expected, see Bolzon *et al.* (1995), the computational effectiveness of the method turns out to depend primarily on whether an enumerative (Judice and Mitra, 1988) or a path-search procedure (Dirkse and Ferris, 1995) is employed to solve the LCP (6). In fact, the enumerative procedure guarantees that all the multiple solutions (if any) are captured but requires long (sometimes prohibitive, depending on the problem size) computational times. The computing times of the genetic algorithm combined with the path-search procedure have been found to be comparable with those of the SQP algorithm in the numerical tests run so far.

6. Identification strategy and some applications

In order to check the proposed identification procedure, a preliminary study has been carried out on the basis of pseudo-experimental data, i.e. of data generated by solving the direct (analysis) problem for a given set of material parameters, which are compared to those provided by the identification process as an accuracy check of it.

A 400×100 mm unnotched specimen has been considered to this purpose, with the following material parameters: E = 14700 MPa, $\nu = 0.1$; $p_c = 1.285$ MPa, $p_b = p_c/3$, $k = h = 42.83$ MPa/mm ($h^* = 21.416$ MPa/mm).

Due to the symmetry, half specimen has been discretised into a fine mesh (930 4-node finite elements) to evaluate the influence matrices **Z**, **G** and **R**. Tractions and displacement discontinuities on the symmetry axis, which coincides with the discontinuity locus Γ_p, have been modelled so that they depend on 10-component vectors **P** and **W** of generalized variables, respectively. Regular softening behaviour with no snap-back instability, see curve (a) of Figure 2, results from the selected material and geometric characteristics.

The main questions to be answered by preliminary numerical tests were formulated as follows: 1-2) which and how many "loading situations" are to be subjected to (optical) measurement for a successful identification of the four parameters in the cohesive model; 3) how many points should be monitored on the specimen surface for each "loading situation"; 4) which displacement component (either the horizontal or the vertical) is better to monitor. One "loading situation" is represented by one point in the load-displacement curve (one circled point of Figure 2) and is characterised by a corresponding distribution of opening displacements on the interface Γ_p.

An answer to question(s) 1-2) arises from the following considerations.

i) Clearly, only situations on the tail of the load-displacement curve correspond to well developed crack, so that some portion of the discontinuity locus Γ_p reflects the p versus w curve of Figure 1(b) beyond the breakpoint. Hence, useful information can be experimentally obtained about all the fracture parameters to be identified. On the contrary, as already pointed out by Alvaredo and Torrent (1987), the peak load Q_p and the corresponding displacement q_p (see Figure 2) depend, for typical specimen dimensions, on the tensile strength p_c and on the initial slope k of the interface law only. Therefore, these parameters are separately identifiable on the basis of data concerning loading situations near the peak.

ii) It can be expected, and was numerically shown, that data from one loading situation only do not allow an easy convergence toward the (here known) exact values of the parameters to be identified; in fact, in this case a plot of the objective function ω (8) in the space of the fracture parameters, p_c and k, shows a "valley" of minimum points. However, considering more loading situations together increases the dimension of the problem to be minimised, with a much more than proportional increase of the computing time.

iii) As anticipated in Section 3, the special case of a single-branch linear softening law, Equations (3) and (4), gives rise to a discretised problem whose dimension are much smaller than those arising from the bilinear interface law.

The above considerations suggested to uncouple the identification problem. In fact, in a first phase, a single-branch linear softening law was assumed, and a good estimation of the parameters p_c and k was easily obtained from the data of two loading situations near the peak load, with a computing time of few minutes on an HP700 workstation. The results of this first phase were introduced as starting values in the second identification phase. Data from a single loading situation on the tail of the Q versus q plot where used to determine the two further parameters p_b and h and to improve the values of p_c and k (which were not kept fixed). The computing time of the second phase was of few minutes. In each phase, the input data were the horizontal displacements of 48 points, in a equally spaced 6×8 (rows×columns) grid.

To answer the afore formulated question 3), regular grids with different numbers of points were considered in a sequence of comparative identification processes. The results showed that below a critical value of about 20 points, convergence toward the known exact values turns out to be very slow; moreover, the computational time decreased with the increasing number of points, up to the limit of about 40-50 data. However, the number of readings cannot increase much beyond a threshold like this, because the ESPI equipment used for measuring the displacements requires a minimum distance of one point from an other for reliable readings.

Finally, with regard to question 4), as expected, for a given grid, the horizontal displacement components contain much richer information to the present purposes than the vertical ones, and, in fact, led to higher accuracy and faster convergence in the numerical tests which were ran.

The experience gained by dealing with the pseudo-experimental data was exploited to identify the fracture behaviour of an innovative quasi-brittle polymeric composite made of an epoxy matrix with silicon micro-spherical hollow inclusions. The elastic moduli were: E = 3707 MPa, ν = 0.39.

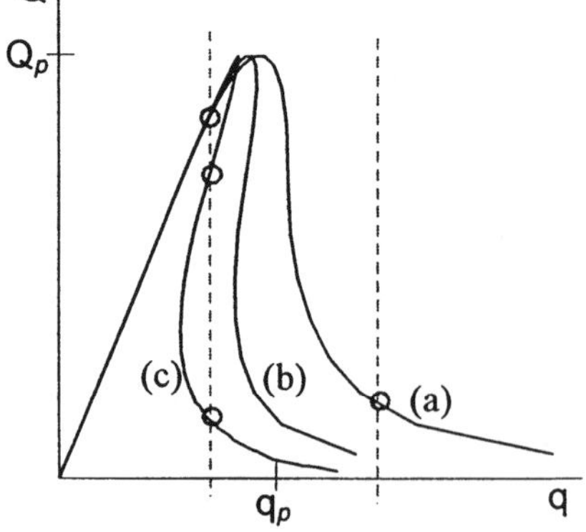

Figure 2. Single-step solution(s) under displacement control, depending on material brittleness.

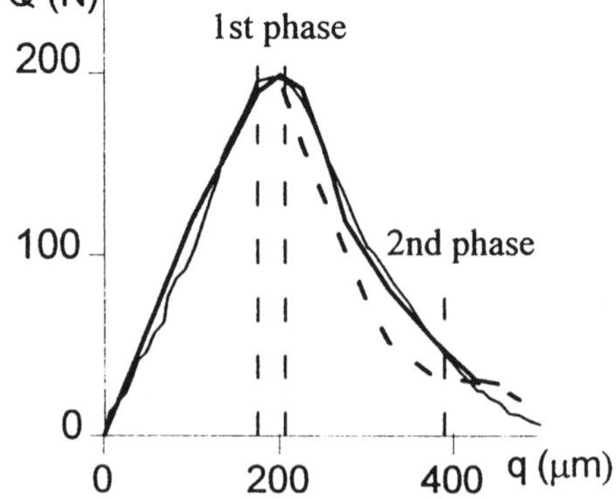

Figure 3. Experimental and identified fracture behaviour of a composite material.

Regularly softening post-peak behaviour (shown by Figure 3, thin line) was observed during 3PB of 107.1×30 mm notched specimen, 15 mm thick, 15 mm notch depth. The horizontal displacements of 30 points in a regular 5×6 grid have been monitored during the experiments.

A fine discretization (1840 4-node finite elements), with 10 nodes on the ligament, was employed in the numerical simulation of the test. A linear softening law was assumed in a first phase. The loading situations used in the first and second phase of the parameter identification process are shown in Figure 3 by vertical dashed lines. In Figure 3 the experimental load-displacement curve is compared to the numerical simulations using, as input fracture parameters, the results of the first (dashed line) and of the second (thick line) identification phase. A reasonable agreement between computed and experimental data is already found with the single-branch linear softening law hypothesised in the first phase, but the bilinear law leads to a much closer fit all along the fracture experiment.

Faster convergence and better results were obtained in this case by the SQP procedure, compared to the GA. This partly depends on the different way the two methods impose the equilibrium constraint in the minimisation of the objective function, but this point deserves further investigations.

7. Conclusions

An indirect identification procedure apt to evaluate the parameters of piecewise-linear cohesive crack models with breakpoint has been discussed, and its ability to provide reliable results has been assessed by numerical tests. The procedure uses, as main experimental data, the horizontal displacements (measured by optical means) of some points on the surface of the specimens in standard three-point bending tests, and a mathematical model for the computer simulation of the tests in the format of a linear complementarity problem (LCP).

The fracture parameters which minimise the discrepancy between measured and computed displacements have been comparatively identified both by a (gradient) sequential quadratic programming (SQP) and by a (direct search) genetic algorithm (GA). The computational experience gained so far, has shown that these two alternative methods are almost equivalent when applied to pseudo-experimental data, which are not affected by measure errors, but faster convergence and better results were found by the SQP procedure with real data.

Acknowledgements — The authors wish to thank Prof. Sergio Bittanti (Politecnico of Milan) and his co-workers for friendly advices on the genetic algorithms. Thanks are also due to Dr. Renata Morbiducci and Vincenzo Anelli for performing the experimental tests on the composite material. A so called "MURST 40 %" research grant on mechanics of materials is also acknowledged.

References

Alvaredo, A.M. and Torrent, R.J. (1987) The effect of the shape of the strain-softening diagram on the bearing capacity of concrete beams. *Mat. Struct.* 20, 448-454.

Arora, J.S., Elwakeil, O.A., Chachande, A.I. and Hsieh, C.C. (1995) Global optimization methods for engineering applications: a review. *Struct. Optim.* 9, 137-159.

Bittanti, S., Maier, G. and Nappi, A. (1983) Inverse problems in structural elasto-plasticity: a Kalman filter approach. In *Plasticity Today* (A. Sawczuk, G. Bianchi eds.), Elsevier Applied Science Publishers, London.

Bolzon, G. (1996) Hybrid finite element appoach to quasi-brittle fracture. *Computer & Structures* 60, 733-741.

Bolzon, G., Maier, G. and Novati, G. (1994) Some aspects of quasi-brittle fracture analysis as a linear complementarity problem. In *Fracture and Damage in Quasi-brittle Structures* (Z.P. Bazant, Z. Bittnar, M. Jirásek and J. Mazars eds.), E&FN Spon, London, pp. 159-174.

Bolzon, G., Maier, G. and Tin-Loi, F. (1995) Holonomic and nonholonomic simulations of quasi-brittle fracture: a comparative study of mathematical programming approaches. In *Fracture Mechanics of Concrete Structures* (F.H. Wittman ed.), Freiburg, 885-898.

Davis, L.D. (1991) *Handbook of Genetic Algorithms*, Van Nostrand Reinhold, New York.

Dirkse, S.P. and Ferris, M.C. (1995) The PATH solver: a non-monotone stabilization scheme for mixed complementarity problems. *Optim. Meth. & Software* 5, 123-156.

Facchini M. and Zanetta P. (1996) ESPI in-plane system applied to the evaluation of masonry mechanical characteristics. *J. Measurement Sci. and Techn.*

Goldberg, D.E. (1989) *Genetic algorithms in search, optimization and machine learning.* Addison-Wesley.

Goretti, A., Valente, C. and Perno, S. (1995) Indirect determination of the fracture properties through three point bending tests. In *Computational Methods and Experimental Measurements* (C.A. Brebbia ed.), Berlin, 611-618.

Guinea, G.V. (1995) Modelling the fracture of concrete: the cohesive crack. *Mat. Struct.* 28, 187-194.

Guinea, G.V., Planas, J. and Elices, M. (1994) A general bilinear fit for the softening curve of concrete. *Mat. Struct.* 27, 99-105.

Hillerborg, A. (1985) Results of three comparative test series for determining the fracture energy G_f of concrete. *Mat. Struct.* 18, 407-413.

Hillerborg, A. (1991) Application of the fictitious crack model to different types of materials. *Int. J. Fracture* 51, 95-102.

Judice, J.J. and Mitra, G. (1988) An enumerative method for the solution of linear complementarity problems. *Eur. J. Operat. Res.* 36, 122-128.

Luo, Z.-Q., Pang, J.-S. and Ralph, D. (1996) *Mathematical programs with equilibrium constraints.* Cambridge University Press.

Maier, G. (1981) Inverse problem in engineering plasticity: a quadratic programming approach. *Atti della Accademia Nazionale dei Lincei, Serie VIII*, 70, 203-209.

Maier, G. Novati, G. and Cen, Z. (1993) Symmetric Galerkin boundary element method for quasi-brittle fracture and frictional contact problems. *Comp. Mech.* 13, 74-89.

Maier, G., Giannessi, F. and Nappi, A. (1982) Identification of yield limits by mathematical programming. *Engrg Structures* 4, 86-98.

Matlab (1996) *User's guide and optimization toolbox user's guide*. The Math Works Inc., USA.

Nanakorn, P. and Horii, H. (1996) Back analysis of tension softening relationship of concrete. *J. Materials, Concrete Structure, Pavements* 32, 265-275.

Planas, J. and Elices, M. (1991) Nonlinear fracture of cohesive materials. *Int. J. Fracture* 51, 139-157.

RILEM Draft Recommendation (1985) Determination of the fracture energy of mortar and concrete by means of three-point bend tests on notched beams. *Mat. Struct.* 18, 285-290.

Wittmann, F.H. and Hu, X. (1991) Fracture process zone in cementitious materials. *Int. J. Fracture* 51, 3-18.

MATERIAL PROPERTY ASSESSMENT IN CRACKED REINFORCED CONCRETE BY DYNAMIC SYSTEM IDENTIFICATION

J.M. NDAMBI, J. DE VISSCHER, G. VAN VINCKEROY, W.P. DEWILDE
Free University of Brussels, Structural Analysis Dept.
Pleinlaan 2, B-1050 Brussels, Belgium.
J. VANTOMME
Royal Military Academy, Civil Engineering Dept.
Renaissance Avenue, 30, B-1000 Brussels, Belgium
B. PEETERS, M.A. WAHAB, G. DE ROECK
K.U.Leuven, Civil Engineering Dept.,
W. De Croylaan 2, B-3000, Heverlee, Belgium

Abstract

This paper reports on the use of dynamic system characteristics (DSC'is) for monitoring of the structural integrity of structures in civil engineering. The experiments are made on two reinforced concrete beams. These beams, with a length of 6 m, are suspended from a frame by means of very elastic springs, in order to realize free boundary conditions during dynamic measurements. Cracks are induced by loading the mid-section of the beams in seven different steps. After each step, a complete Experimental Modal Analysis is performed to determine the DSC's of the beams. Their evolution allows to draw conclusion about the suitability of the experimental techniques. This investigation is the first step in the development of a DSC based technique for monitoring damage in large concrete structures. It is shown here that increasing crack damage results in measurable frequency shifts, increasing curvature in the cracked zone and a shift of nodal points towards the centre of the beams.
A finite element model (FEM) is also generated for the test structure for the purpose of comparing with the experimental results.

1. Introduction

Any crack or other damage in concrete structures, whether local or global, is associated with structural modifications. These modifications can be observed through changes in modal parameters, i.e., modal frequencies, modal damping values, mode shapes and their derivatives, so that the use of dynamic system characteristics (DSC'is) for damage assessment is very promising. Changes also occur in some of structural parameters: mass, damping, stiffness and flexibility matrices of the structure.

H. Sol and C.W.J. Oomens (eds.), Material Identification Using Mixed Numerical Experimental Methods, 223-232.

In recent years, effort has been made to use modal parameters to detect structural damage in concrete structures. This idea is already accepted, but the practical exploitation is rather difficult, because of the drawing of the correct correlation between changes of modal parameters and the nature of damage, its location and its severity.
In the domain of fiber-reinforced plastics, some researchers have shown that damage in these materials can be detected from the decrease of modal frequencies and the increase or decrease of damping **[A.K. Pandey, 1994]**. The aim of this research is to investigate for the case of concrete structures. Which modal parameter can be used to detect damage, its location and its severity?
In this paper, the influence of progressive cracking on modal frequencies, modal damping, and mode shapes is investigated.
Another approach for damage location is based on mixed numerical / experimental analysis. By tuning the parameters in a numerical model in such a way that the numerically calculated dynamic data fit experimental data, it should be possible to quantify and locate damage in the structure. The main difficulties encountered in such an approach are twofold: the first difficulty is to determine a set of physically relevant variables to describe damage, the second is the control of the influence of all parameters influencing the dynamic behaviour of the structure, besides the damage itself. Obviously, when the influence of those parameters is not well understood, wrong conclusions could be drawn concerning the actual damage.

2. Experiments

Two different types of experiments are combined: A static loading test and dynamic measurements.

2.1. DESCRIPTION OF THE TEST BEAM

Tests are prepared and realized in the laboratory of the Civil Engineering Department of the K.U.Leuven. Two test beams were manufactured using the same design parameters. The test specimens are reinforced concrete beams, with 6 m length, 25 cm width, and 20 cm height. The beams are reinforced with 6 longitudinal steel (S500) bars with 16 mm diameter and transverse reinforcement consisting of stirrups with 8 mm diameter each 20 cm. Figure 1 shows the cross section of the beam.

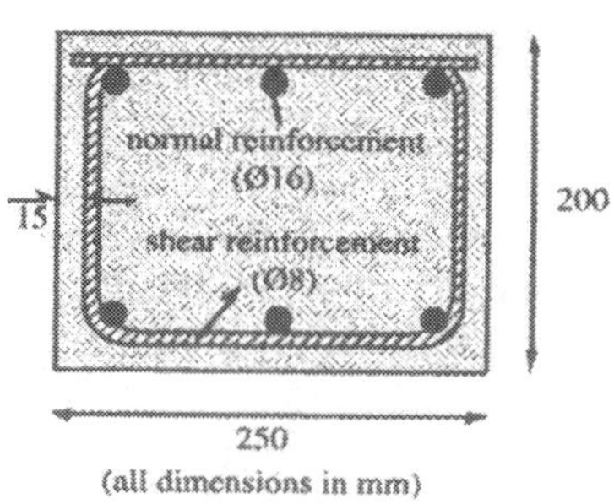

Figure 1. Cross section of the beam

The beams are designed in order to reach the following objectives:
- The first eigenfrequency have to be comparable with the lowest eigenfrequency generally encountered in civil engineering structures (e.g. bridges).
- The cross section is chosen rectangular to avoid coupling effect between horizontal and vertical bending modes.
- The reinforcement ratio is chosen to have a sufficient difference between the first crack moment and the failure moment of the beam.
- Transverse reinforcement ensures that failure occurs in bending.

2.2. MEASUREMENT SYSTEM

2.2.1. Dynamic measurements

Dynamic parameters are determined by Experimental Modal Analysis which is used to describe the dynamic behaviour of a mechanical system in terms of its vibrations modes. A serie of Frequency Response Functions (FRF) are measured by means of the Fast Fourier Transform (FFT) analyzer. The curve fit on these functions permit to determine all modal parameters.
- For the excitation, two systems are used: the first one is the impact loading with hammer that include an integral piezoelectric sensor to measure the applied force. A hammer type Dytran (5802) S/N 1300 with sensitivity ± 0.92 mV/Lb-F is used in the lower range of excitation frequencies, and a hammer type Dytran (5801A4) S/N 1394 with sensitivity ± 9.9 mV/Lb-F is used for the higher range. The second system uses an electromagnetic shaker with force sensor PCB S/N 13422 with sensitivity ± 52 mV/Lb-F. The shaker is driven by the signal generated by the FFT analyzer output channel. The periodic random signal is used.
- The vibration responses of the structure are measured by means of accelerometers: type Dytran and PCB with sensibility ± 100 mV/g.
- The data acquisition system consists of a dual FFT analyzer ONO SOKKI and a personal computer. All modal parameters are calculated with the Rational Polynomial fit-method using the ENTEK-EASY software.

2.2.2. Static loading test

The static loading is applied by means of an hydraulic system. The vertical displacements are measured with the linear variation displacement transducers (LVDT).

3. Experiments set-up and procedures

(a) The static loading test is a three-bending quasi-static load with a span of 3600 mm. (Figure 2)
The load P is applied in the middle of the beam, to locate the cracked area in the centre of the beam. The beam is simply supported. The position of supports are calculated in view to minimize the bending moment due to own weight of the beam. Damage is induced in seven different steps during which accumulation of cracks is observed.

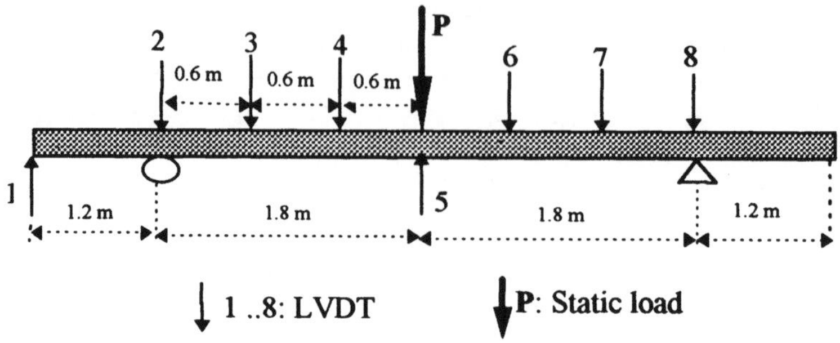

Figure 2. Static test configuration

Table 1 resumes the static loading test programme.

Table 1. Static load steps

Date (test)	Load step	Load (kN)
96-30-10	ref	0
96-13-11	1	8
96-19-11	2	15
96-21-11	3	24
96-02-12	4	32
96-05-12	5	40
96-12-12	6	50
96-19-12	7	56

During the static loading test, some measurements are made:
- Force displacement measurements (LVDT)
- Strain measurements
- Visual inspection of the crack pattern

(b) After each static loading step, dynamic measurements are performed to study the evolution of modal parameters. Figure 3 shows the dynamic test configuration.

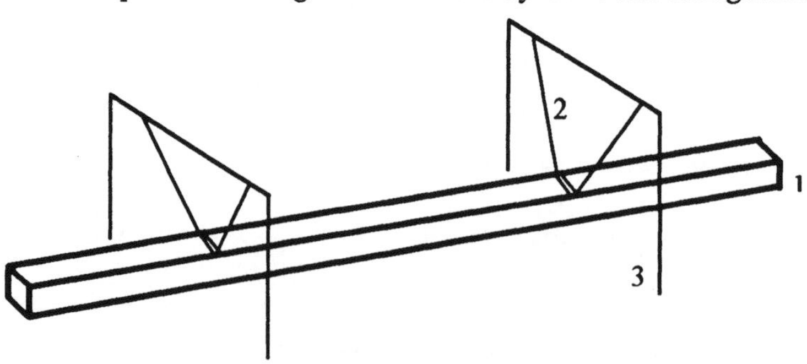

Figure 3. Dynamic test configuration
1: Beam; 2: Springs; 3: Frame

The beams were suspended by means of four very elastic springs attached at the theoretical nodal points of the fundamental bending mode of the beam (1.344 m to each end) in free free conditions[**S. Thimoshenko, 1974**]. These elastic springs are used to reduce the influence of the environment and the suspension system on the dynamic behaviour of the beam..

4. Experimental Results

4.1. STATIC LOADING TEST

Only the results of the second beam are presented.
The static load is applied in seven steps. In the first step, the maximum force is limited to 8 kN (before the occurrence of the first visible crack). The failure of the beam occurred at 56 kN. Figure 4 shows the force vs. displacement curve of the seven static loading steps, measured at the mid-section of the beam.

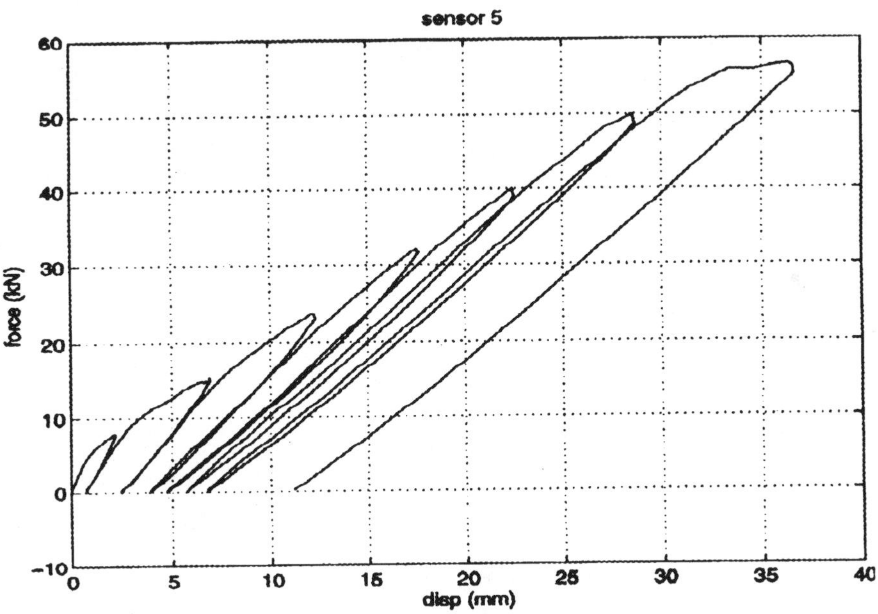

Figure 4. Force-displacement curve at the mid-section.

The first visible crack was observed in the vicinity of the mid-section of the beam. During the static load programme, other cracks developed towards the supports. Failure occurred when a crack at the mid-section of the beam had grown over the entire section.

4.2. DYNAMIC TEST

The evolution of modal parameters is described in table 2, figure 6 and figure 7.

The analyzed frequency range is 0-1000 Hz in order to study the first fourteen vibration modes of the beam. Ten bending and four torsion modes are identified. Figure 5 shows some experimental mode shapes of the beam.

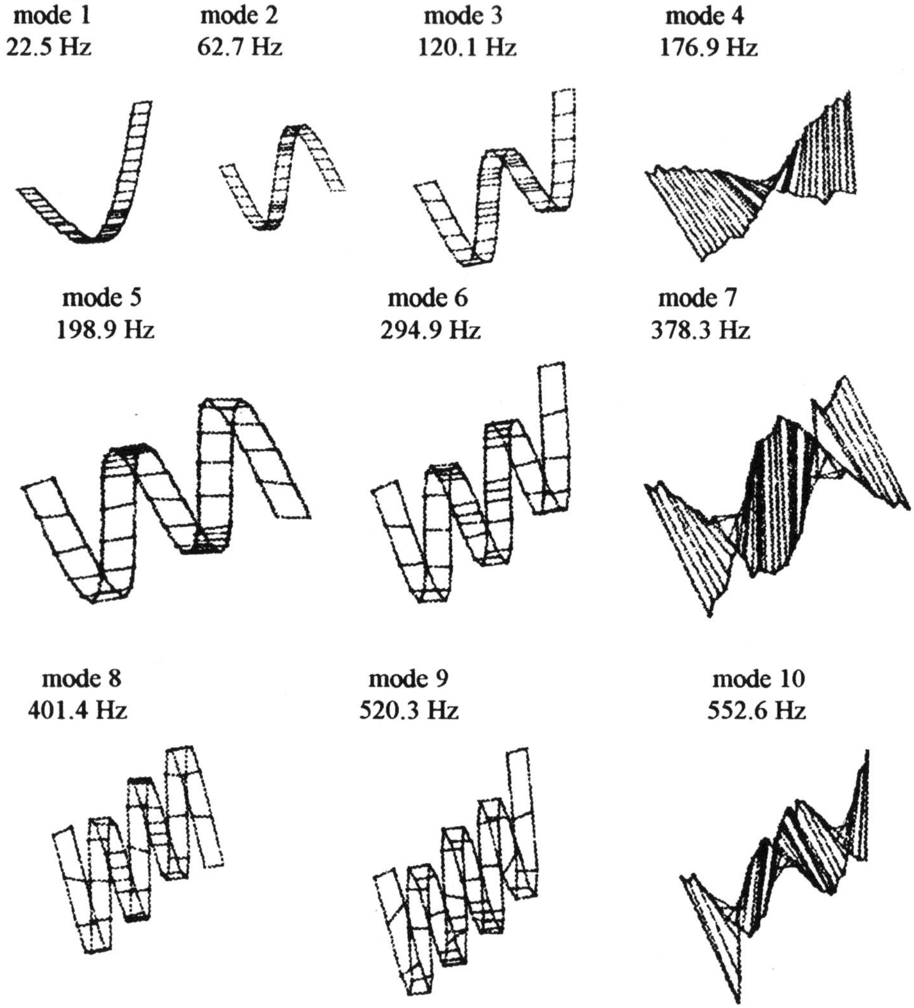

Figure 5: Experimental mode shapes for the first 10 vibration modes of the beam. The modal frequencies correspond to the undamaged state.

The figure 6 gives the relative variation of the eigenfrequencies of the 10 first modal frequencies of the beam.

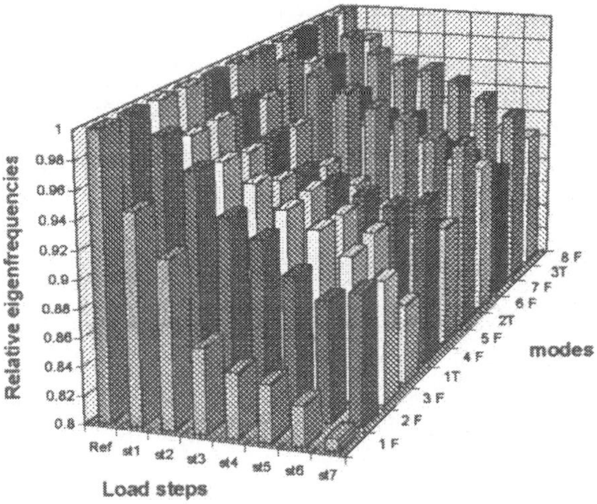

Figure 6. Relative variation of the eigenfrequencies
B: bending mode; T: torsion mode

Table 2 gives the evolution of the 10 first modal damping ratio's of the beam.

Table 2. Evolution of modal damping ratio's

Static loading step	Modes									
	1 B	2 B	3 B	1 T	4 B	5 B	2 T	6 B	7 B	3 T
ref.	0.32	0.29	0.30	0.61	0.30	0.32	0.33	0.34	0.32	0.32
1	0.47	0.31	0.29	0.62	0.31	0.33	0.30	0.33	0.35	0.29
2	0.59	0.38	0.52	0.45	0.53	0.58	0.37	0.72	0.69	0.44
3	0.66	0.54	0.54	0.51	0.68	0.84	0.36	0.70	0.67	0.47
4	0.79	0.37	0.52	0.53	0.51	0.63	0.37	0.43	0.38	0.36
5	0.86	0.66	0.64	0.38	0.66	0.88	0.48	0.39	0.67	0.38
6	0.75	0.61	0.44	0.39	0.63	0.96	0.45	0.47	0.53	0.56
7	0.80	0.55	0.723	0.68	0.56	0.57	0.52	0.59	0.57	0.41

B: bending mode; T: torsion mode

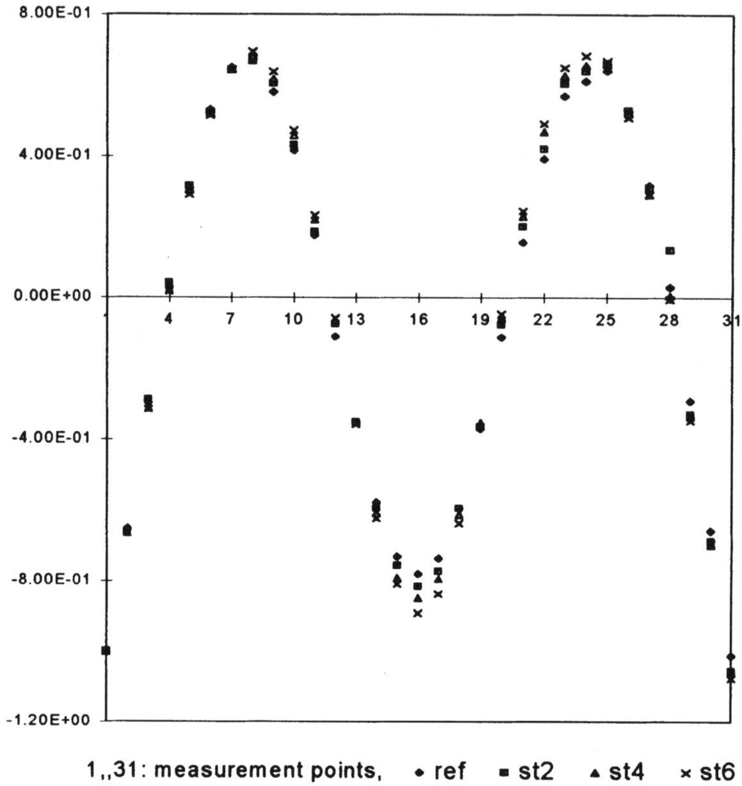

Figure 7: Change of the shape of the third bending mode

These tables are susceptible to some comments:

- It appears from these results that modal frequencies decrease with damage accumulation. This phenomenon can be explained by the fact that when cracks occur in the concrete structure, its EI stiffness decrease and the modal frequencies vary in the same way **[B.D. Zakic, 1996]**.

The first bending mode is more affected than the other by the accumulation of damage, because of its maximum amplitude of vibration in the cracked area. In general, symmetric modes of vibration are more affected than the asymmetric ones. After the failure of the beam, a sudden drop of modal frequencies is observed in the symmetric modes. These observations permit to say that the modal frequencies can be considered as a good parameter for detecting damage in concrete structures.

- By observing the results on the modal damping ratio's, it is difficult to establish any relationship between its evolution and the cracking accumulation. This because of the complex dependence of this parameter. Modal damping is very sensitive and very difficult to measure. The relevance on the result depends partly on the measurement technique, partly on the fitting method (quality of frequency response functions or time histories), and partly on the environment. Consequently, modal damping is not at the

moment a suitable parameter for detecting damage in concrete structure. The improvement of the measurement technique should be encouraged in the future.
- The figure 7 shows the change of mode shape of the third bending mode. Only the shapes that correspond to the reference, the second, the fourth and the sixth static loading steps are represented. It can be observed that the nodal points shift towards the damage area and the relative modal amplitudes increase in this area.

5. Numerical simulations

Frequency shift gives a global indication of the presence of damage, while the variations of mode shapes and especially curvature mode shapes give an indication of the location of damage. All these results follow from a purely experimental analysis of the structure.
A finite element model of the beam has been developed by means of F.E. program SAMCEF. The model comprises 10320 elements, from which 8640 are first degree volume elements representing the concrete, and 1680 are first degree beam elements to represent the steel reinforcement (normal reinforcement and stirrups). The data under study are the eigenfrequencies for the dynamic analysis and the maximum vertical displacement for the static analysis. Influence of mechanical and physical parameters on these data are thoroughly investigated before introducing any damage.
The need of such preliminary analysis arises from the fact that the properties to be introduced in the finite element model are obtained from experiments on specimens (i.e. cement cylinders) different in nature than the real structure (beam). Another point to take into account are the numerical approximations inherent to the model: Type finite element, model geometry, etc...
The eigenferquencies obtained from the finite element model are presented in the table 3.

Table 3. Eigenfrequencies for F.E. model and experiments in Hz

	Experimental results	F.E. Model Case a	F.E. Model Case b	F.E. Model Case c
Modes		E_a=218E9 Pa E_b=35.6E9 Pa NU_a=0.28 NU_b=0.2	E_a=218E9 Pa E_b=32.2E9 Pa NU_a=0.28 NU_b=0.2	E_a=218E9 Pa E_b=32.2E9 Pa NU_a=0.28 NU_b=0.15
1 B	22.50	23.12	22.74	22.73
2 B	62.70	63.21	62.17	62.17
3 B	120.1	122.4	120.5	120.5
1 T	176.9	171.2	172.5	176.2
4 B	198.9	199.4	196.2	196.3
5 B	294.9	292.4	287.8	288.2

E_a, NU_a: E-modulus and Poisson coefficient of steel.
E_b, NU_b: E-modulus and Poisson coefficient of concrete

In this table, the experimental eigenfrequencies are given in an undamaged state of the beam.

6. Conclusion

It appears from these experiments that:

- The increasing of cracks damage are reflected in modification of modal parameters. The cracks damage results in a measurable frequency shift, increasing curvatures and a shift of nodal points towards the damage area of the beam.
- The modal damping is very sensitive and difficult to measure. It depends on the measurement technique and the fit-method used to extract modal parameters. Possibilities to increase the accuracy of measurements and frequency resolution should be encouraged, because the quality of the FRF and the subsequent curve fit depend on it.

7. Acknowledgement

This work is a part of a research project sponsored by the National Fund for Scientific Research. Its financial support is gratefully acknowledged.

References

Pandey, A.K. (1994) Experimental verification of flexibility difference method for locating damage in structure, Journal of Sound and Vibration, (1995) 184 (2), 311-328.

Zakic, B.D. (1996) Vibration in diagnosis of damages in concrete bridge structures, Proceedings of 2nd RILEM International Conference, Diagnosis of concrete structures, 320-323.

Yiu, W. L.; Coleman, R.E., Zhang, W.X., Bing,X.X., Luo, J., Liao, Q.W. (1996) Modal Analysis of Chinese Standard Structure using Varied Excitation Method, 14 Th. Int. Modal Analysis Conf., Dearborn, Mishigan, 8-15.

Pandey, A.K. Biswas, M. (1990) Damage detection from changes in curvature mode shapes, Journal of Sound and Vibration, (1991) 145(2), 321-332.

Ewins, D. J. (1986) Modal Testing: theory and practice, Brüel & Kjaer.

Toksoy, T, Aktan A.E. (1993) Bridge-condition Assessment by Modal Flexibility, SEM Spring Conference on Experimental Mechanics, Deaborn, MI, 271-278

Abstracts from the presented posters

ELASTIC MODULI IDENTIFICATION OF A COMPOSITE TUBE AND AN EPOXY RESIN PLATE USING BOTH EXPERIMENTAL AND NUMERICAL MODAL ANALYSIS - SENSIBILITY ANALYSIS.

P.AMAT, D.GAY
Laboratoire Genie Mecanique de Toulouse
L.G.T.M./I.U.T.G.M.P. 50, Chemain des Maraîchers 31077 Toulouse

In order to model the mechanical behaviour of a reservoir built in composite materials, we have first performed a study of the elastic moduli of an epoxy resin circular plate and a composite tube. The elastic moduli of the tube and the plate cannot be identified according to standard tensile tests. Therefore we have settled a modal analysis technique which consists of an identification of the free vibration modes. This identification has been done experimentally and numerically by finite element modelling.

We have thus updated the elastic moduli of models and after a few iterations we have obtained a good correlation with the experimental results.

The experimental modal analysis enable the symmetric and antisymmetric vibration modes to be determined. The numerical modal analysis have been performed using axisymmetric, shell and volumetric elements. So we have increased the correlation between the experimental and numerical modal analysis and we have improved the characterisation process accuracy.

H. Sol and C.W.J. Oomens (eds.), Material Identification Using Mixed Numerical Experimental Methods, 235.

SOIL IDENTIFICATION USING DATA PROCESSING OF PENETRATION EXPERIMENTS

D.B.BALASHOV
Institute for Problems in Mechanics of RAS,
101 Vernadskogo avenue, 117526 Moscow

The new indirect method for the determination of the dynamic yield strength for soil is suggested [1]. The yield strength is postulated to be equal to the value of the tangential stresses on the surface of the indentor penetrating the soil. According to the experimental results for soils the dependence of drag force and yield strength on the strain rate is neglected for the velocities greater than the critical value. This assumption permits to describe the stress state in the vicinity of the conic indentor by the constructed solution for the quasistatic self-similar problem on the flow of a rigid-plastic Mises-Schleicher solid over an infinite rigid cone. Unlike the paper [2] the continuous velocity field on the bound of plastic and rigid regions that occurred in the penetration experiments is taken into account. The solution of this modelling problem is reduced to the investigation of ordinary differential equations of the first order. The qualitative analysis of the solution for the stress and velocity fields around the cone with respect to friction regime and cone angle is given. The condition is found under which there exists a solution. Using the inverse method, the *a priori* unknown yield strength is selected in such a way that the solution satisfies experimentally justified assumptions about the penetration process. This method permits to estimate the yield strength using results of a single experiment by penetration of a freely moving conically tipped massive circular rod into the soil with constant velocity. The experiment involves the measurement of the drag force for the moving tip. The previous method [3] is based on the extrapolation of the experimental data and requires a series of experiments. To verify the new method the measurement data of drag force are used in yield strength calculations. Good fitness with known estimations is discovered.

1. Balshov, D.B., Zvolinsky, N.V., Flow of a rigid plastic medium around a cone. Mech. Of Solids., **31**, (1996), 39-45.

2. Durban,D., Fleck, N.A., Singular plastic fields in steady penetration of a rigid cone. J. Appl. Mech., **59**, (1992), 706-710.

3. Bivin, Yu.K., Viktorov,B.Ya. Determination of the dynamic characteristics of the soil using penetration method. Izv. Akad. Nauk, MTT, USSR, **3**, (1980), 105-109.

H. Sol and C.W.J. Oomens (eds.), Material Identification Using Mixed Numerical Experimental Methods, 236.

COMPLETE ELASTIC CHARACTERIZATION OF TRANSVERSELY ISOTROPIC COMPOSITE RODS BY GUIDED STRUCTURAL WAVES

J.VOLLMANN, M.R.PFAFFINGER, J.DUAL

Institute of Mechanics, ETH Zürich, Swiss Federal Institute of Technology
CH-8092 Zürich, Switzerland

The material properties of a carbon fibre reinforced composite rod with circular cross-section are determined by measuring the dispersion relation (wave speed versus frequency) of several axisymmetric wave modes and by comparing it with a theoretical model. The axisymmetric waves are excited by a ring-shaped piezoelectric transducer which is subjected to a broad band pulse. Constructive interferences between P-, SH- and SV-waves and their reflections lead to propagating wave modes. The dispersion relation of these wave modes contains characteristic information about the geometry of the structure and it s material.

New dynamic metrology was developed to measure complex dispersion relations in a frequency range between 1 kHz and 2 Mhz of up to 40 travelling wave modes in one single process. With a heterodyne LASER interferometer, which was partly developed at the institute of Mechanics, ETH Zürich, the radial and the longitudinal displacement of the rod surface is detected at many points along the rod.

The signals are processed with FFT in the time domain. To decompose the complex wave number dependence, a complex spectrum estimation method is used in the space domain. The method, consisting of Linear Prediction and Total Least Squares, although computationally more intensive, demands fewer measurements than FFT and allows a lower signal to noise ratio. It extracts both real and imaginary parts of the wavenumber from the measured data. Finally the parameters (material properties) in an exact model previously developed by Mirsky are systematically varied until an optimal fit with the measured dispersion curves is obtained. In this manner the five independent material constants, describing the elastic behaviour of transversally isotropic materia are determined with high accuracy.

H. Sol and C.W.J. Oomens (eds.), Material Identification Using Mixed Numerical Experimental Methods, 237.

IDENTIFICATION OF MATERIAL PARAMETERS FOR MEASUREMENT ERROR MODELS VIA MULTIPLIER METHODS

THOMAS FLÖCK
Institut für Mechanik
TU Chemnitz-Zwickau, Chemnitz

Introduced is a linearized Multiplier Method which is especially constructed for the determination of material parameters which is based on measurement error affected slopes and yield stresses for initial yield curves. At issue is especially the calculation of error propagation, which is supposed to be based on the properties of normal distribution. For this purpose three kinds of generalized least-squares are considered. In this context it is necessary to evaluate the true stresses and the corresponding slopes to determine exact material parameters which will fulfill the initial yield function in 4 points in the principal stress plane.

Thus this article is dedicated to both statistical points of view and fit methods of nonlinear constrained optimization to evaluate material parameters with consideration of erroneous data. Furthermore for the sake of well-condition a Tikhonov-Phillips regularization technique is required, which will lead to quasi optimal statements. As for the optimization method a approach due to Rockafellar and a updating technique corresponding to Bertsekas is applied.

This lecture thus focuses upon elasto-plastic material behaviour under the assumptions of small deformations and the rule of normality. Furthermore it is supposed that the measurement errors are founded only in the testing setup and not in the material itself.

H. Sol and C.W.J. Oomens (eds.), Material Identification Using Mixed Numerical Experimental Methods, 238.

FIBRE/MATRIX INTERFACE FRACTURE TOUGHNESS IDENTIFICATION IN SINGLE FIBRE FRAGMENTATION TEST

ROBERTS JOFFE, JANIS VARNA
Dept. Of Polymer Engineering
Luleå University of Technology, S-971 87 Luleå, Sweden.

The single fibre fragmentation test is perhaps the most frequently used micro mechanical test method for characterization of fibre/matrix interface properties. Usually, the data reduction is based on a interface shear strength approach and the data reduction is performed using rather simplified stress analyses. However, the exact solution at the fragment end contains shear stress singularity and in coupling the use of maximum stress criteria is highly questionable. As a result, obtained values of shear strength depend on the model employed and may be used only for qualitative comparison of interfaces.
In the presented study a more advanced approach to interface failure analysis is performed. The failure criterion is based on energy considerations: total energy released during debond propagation is equal to the work needed to create a new crack surface. This work is proportional to interface fracture toughness and the results are expressed in terms of G_C.

The fracture toughness identification problem is solved in the following way. The stress state is needed for two debond sizes in order to calculate the released strain energy when debond grows from size α to $\alpha + \delta$. Released energy is calculated directly as the difference given by the specimen strain energy for two considered debond sizes. The method requires an accurate stress state in the whole domain and calculations are performed using FEM.
The energy failure criterion gives the relationship between debond size and applied far field stress. The predicted curves depend on the unknown value of G_C that may be determined by comparing predictions with test data.

The experimental information needed in this approach is the length of the debonded part along the fibre versus load. Fragmentation tests were performed on carbon and glass fibres in epoxy matrix for varying fibre surface treatment (interface fracture toughness).

Combining test data with model predictions the interface fracture toughness was determined. Results show a very strong dependence of obtained G_C on fibre surface treatment.

H. Sol and C.W.J. Oomens (eds.), Material Identification Using Mixed Numerical Experimental Methods, 239.

A MIXED NUMERICAL AND EXPERIMENTAL APPROACH FOR THE IDENTIFICATION OF ELASTIC PLASTIC RELATIONS IN WIDE RANGE OF STRAINS, STRAIN-RATES AND TEMPERATURE

F. LLORCA, A. JUANICOTENA
Commissariat ál'Energie Atomique, CEA/Valduc,
CEA/B3 BP12, 91680 Bruyéres-de-Châtel, France

The aim of our studies is to use constitutive laws for the simulation of the response of complex structures which are loaded under high dynamic conditions. The ability of constitutive models to describe material behaviour is a function of the mathematical formulation and associated parameters. So, under these conditions, the identification of realistic coefficients must be obtained from the widest range of strains, strain-rates and temperature.

In our presentation, experimental data are comprised of quasi-static tests and dynamic Hopkinson pressure bar tests. Then, mathematical techniques allow to calculate the coefficients of the many constitutive relations selected for further numerical simulations. After this calculation, other experiments are needed for a first validation in the same range of loading.

Now the ability of constitutive models to simulate the material response to complex loadings is the next step of our work. In our opinion, complex loadings mean non homogenous strain and strain-rate fields, 2D-3D geometrical effects, influences of connex elements, ... To verify the validity of the evaluated constitutive relations under such loadings, specific experiments have to be realized and their associated numerical simulations. For us, Taylor tests and expanding ring/sphere tests allow to obtain a first level of complexity in order to push the limits of models validity.

To illustrate our work, we propose results concerning OFHC copper.

H. Sol and C.W.J. Oomens (eds.), Material Identification Using Mixed Numerical Experimental Methods, 240.

SPRINGER NATURE

GPSR Compliance

The European Union's (EU) General Product Safety Regulation (GPSR) is a set of rules that requires consumer products to be safe and our obligations to ensure this.

If you have any concerns about our products, you can contact us on ProductSafety@springernature.com

In case Publisher is established outside the EU, the EU authorized representative is:

Springer Nature Customer Service Center GmbH
Europaplatz 3
69115 Heidelberg, Germany

Papier | Fördert gute Waldnutzung
FSC® C083411

Kolibri 360 GmbH
Ferdinand-Jühlke-Straße 7
99095 Erfurt, Deutschland
produktsicherheit@kolibri360.de